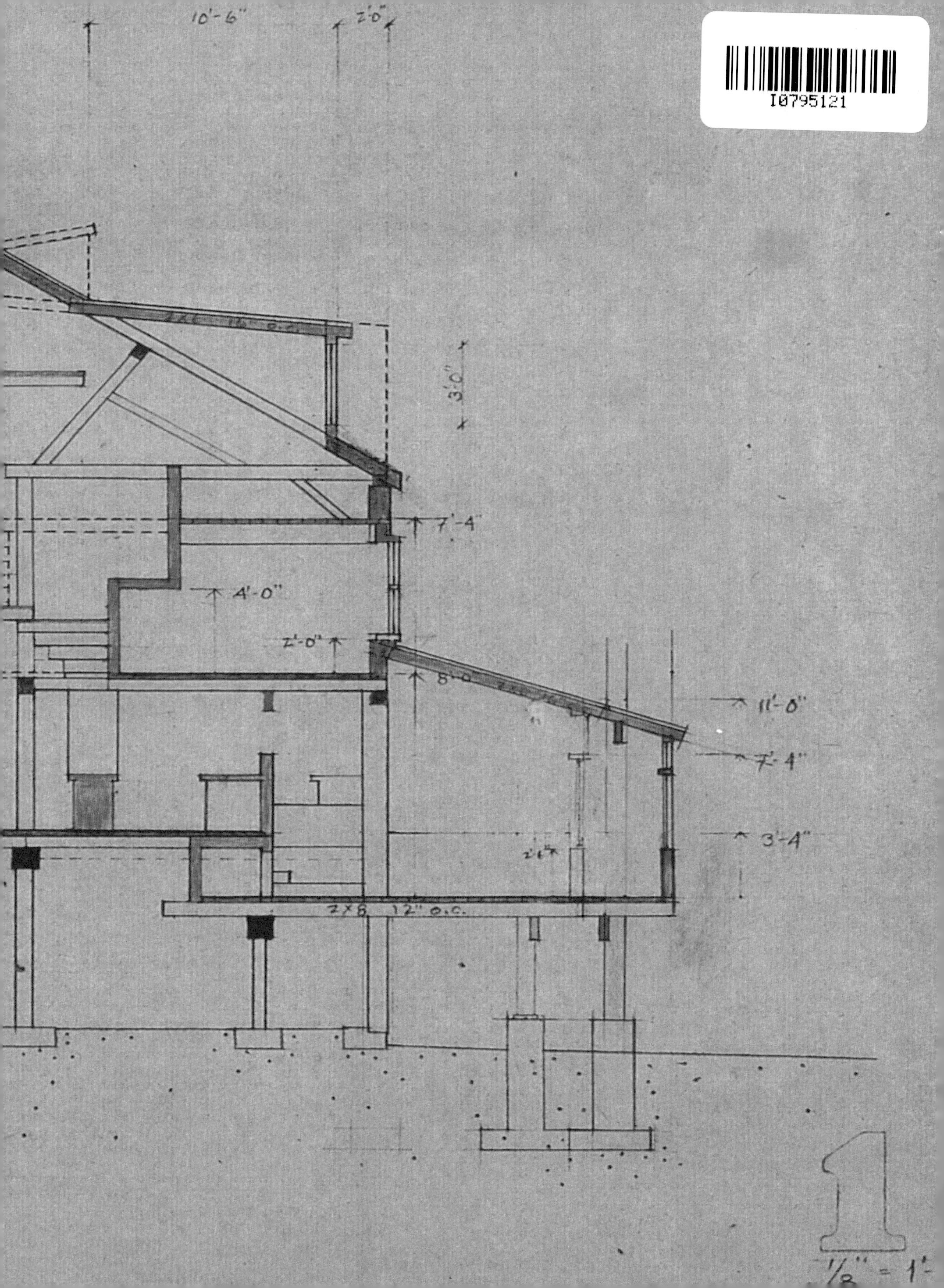
10'-6"
2'-0"
3'-0"
7'-4"
4'-0"
2'-0"
8'-0"
11'-0"
7'-4"
3'-4"
2x8 12" o.c.
1
1/8" = 1'-

Yesterday's Structures: TODAY'S HOMES

Yesterday's Structures:

TODAY'S HOMES

Lucy D. Rosenfeld

Schiffer Publishing Ltd

4880 Lower Valley Road, Atglen, PA 19310 USA

Cover credits:
Front cover: Exterior view today of converted water cistern: the cylindrical structure still has the original decorative cornice and hexagonal tower; updated structure has a new trellised patio, new windows, double-height entry, and newly stuccoed exterior in warm yellow-orange tone. Peter de Bretteville, Architect

Inside front flap: "Before" picture of abandoned cistern (seen on cover) on a ridge above New Haven.

Back cover: Interior of living room has original hand-hewn beams and columns, soaring overhead space, and rustic elegance. Peter Kurt Woerner, Architect.

Inside back flap: The barn frame, with its original rusted tin roof, being moved by truck to its new, larger foundation atop a hill nearby.

For additional front matter credits please see P. 174

Library of Congress Cataloging-in-publication Data
Rosenfeld, Lucy D., 1939-
Yesterday's Structures: Today's Homes / Lucy D. Rosenfeld.
p. cm.
ISBN: 0-7643-1014-3
1. Dwellings--United States--Design and construction Case studies. 2. Dwellings--Great Britain--Design and construction Case studies. 3. Buildings--Remodeling for other use--United States Case studies. 4. Buildings--Remodeling for other use--Breat Britain Case studies.
I. Title
TH4809.U6R67 2000
690' .8--dc21 99-41801
CIP

Design by Blair Loughrey
Type set in Swiss 721 Blk Out Bt/ ZapfHumanist Dm BT heading font/text font ZapfHumanist BT

ISBN: 0-7643-1014-3
Printed in China

Published by Schiffer Publishing Ltd.
4880 Lower Valley Road
Atglen, PA 19310
Phone: (610) 593-1777; Fax: (610) 593-2002
E-mail: Schifferbk@aol.com
Please visit our web site catalog at
WWW.SCHIFFERBOOKS.COM
or write for a free catalog.
This book may be purchased from the publisher.
Please include $3.95 for shipping.

In Europe, Schiffer books are distributed by
Bushwood Books
6 Marksbury Rd.
Kew Gardens
Surrey TW9 4JF England
Phone: 44 (0)208 392-8585; Fax: 44 (0)208 392-9876
E-mail: Bushwd@aol.com

Please try your bookstore first.

We are interested in hearing from authors with book ideas on related subjects.

CONTENTS

I would like to thank the thirty-eight architects who contributed their adaptive reuse projects to this book, as well as the individual designers and do-it-yourselfers who converted their own buildings. These people of imagination have made the book a joy to write. Several architects have been particularly helpful in guiding my hand; I would like to acknowledge the advice of Ray Stainback of Atlanta, Bernard Rothzeid of New York, and Troy West of Newark. My search for interesting projects was aided (via car, telephone, and Internet) by the enthusiastic help of Cecily Morse, Helen Bruner, Theresa Wyman, Deborah Bigelow, and Guy Ladd Frost. Finally, I would like to thank my special and always most faithful reader and supportive critic, my husband Peter.

ACKNOWLEDGMENTS

INTRODUCTION

Adaptive Reuse—the term that describes the conversion of existing buildings from one use to another—represents the intersection of two very contemporary trends. The preservation of our architectural past is finally being taken seriously—not just by preservationists—but by the general public. Recognizing that our cities and villages have an identity that is in large part created by their older buildings and their neighborhoods, planners, architects, developers, and even do-it-yourself individuals are choosing to save buildings that would have been torn down just a few years ago.

The trend toward preserving the architectural past has fortunately come at the same time as the desire of many home owners to live in larger, unpartitioned spaces, and for many architects to turn to creating homes out of older, well-constructed buildings. Some of the most interesting architecture now is being designed to take advantage of the solid structures, over-size windows, high ceilings, and unusual layouts of buildings that were designed to be public spaces (or animal spaces), rather than houses. The desire for open space and loft-style living has made such conversions popular in many parts of the country—from downtown city centers that had long been blighted by abandoned buildings, to rural places where old buildings had become obsolete in villages or on farms.

Obsolete churches, schools, factories, barns, warehouses—these are some of the great and inviting spaces that attract today's adaptive reuse. (Abandoned churches are so popular as candidates for reuse that they are snapped up by buyers all over the country.) But these are not the only candidates for conversion. In this collection, you'll also find other, wildly dissimilar buildings that imaginative people have transformed—a missile silo, a lighthouse, a cistern, a bank, a stable, a power station, a brewery, a railroad station, and even a giant gas tank.

Of course, adaptive reuse can have a variety of objectives. This collection is limited to conversions from spaces where nobody lived, to residential space—either for one family, or for many. In the first half of the book, imaginative transformations of spaces into single family homes (and live/work studios) are presented. In the second section multi-unit housing—often with mixed-use space as well—shows how well large older buildings convert to new uses. In all of them, architects and owners, preservationists and developers, and simply ingenious people who enjoy the serendipitous finds of scavenging and recycling the past have discovered the thrill of making something fine out of abandoned, dilapidated, or obsolete buildings.

In most of these renovations the architectural features of the original building have been preserved whenever possible. For one of the most attractive elements of adaptive reuse is the combination of old, often elegant architectural features with contemporary design and taste. In some conversions, only the historic exterior or façade remains to remind us of its past, while the interior is new and modern. Conversely, in others it is the historic architectural detail of the interior that suggests the former life of the building, while the exterior is a new envelope. In occasional projects very little needs to be done; old public school buildings, for example, were soundly constructed and often require only some interior retrofitting to become viable as apartment houses. Some projects form the architectural center for much larger additions; others occupy only part of a massive space around them.

A few of these retrofitted buildings have been labors of love—taking up to twenty-five years of on-going construction by architects or owners who fine-tune their plans and do their own building. Other, more commercial projects, are completed and put up for sale within a short time—some of them are now among the most desirable spaces available in their regions. Some have been turned—with the help of government and other subsidies—into affordable housing. A few are still in the process of construction into housing. Some have been converted on the proverbial shoestring with salvaged materials and do-it-yourself labor. (Since the costs of adaptive reuse vary so completely from project to project, it would be impossible to discuss here the greatly differing financial elements.)

The projects chosen represent twenty-five states and Great Britain. The architects and the individuals who worked on them are credited on each page; you'll find the names and addresses of the architects listed in the back of the book as well. The architects and designers chosen range from the most established to the young, less well known, to individuals with a taste for renovation—assuring a wide variety of styles. Over two hundred and fifty architects were contacted through recommendations, AIA lists, and published materials.

These projects were selected because they seemed to represent the most interesting of adaptive reuse projects—needless to say, not all such projects are successful. (The balanced conjunction of old and new is not automatically achieved, and there have been plenty of aesthetic failures in such conversions.) While taste will differ on the projects included, each of them has something different to say, and all of them have helped to preserve our architectural past—often in very unexpected ways.

I hope this book will be used and enjoyed in an equally wide variety of ways: for inspiration and new ideas, for the sheer pleasure of looking at before and after pictures of the most imaginative reworkings, for finding an architect to do such a project for you, or finally, because you—like me—have an incurable taste for updated buildings of long ago.

PART 1

CONVERSIONS TO SINGLE FAMILY HOMES

FROM AN URBAN POWER STATION TO A DISTINCTIVE CONTEMPORARY RESIDENCE

CHAPTER ONE

Michael Rotondi and Clark Stevens of RoTo Architects

Overlooking a bustling freight train yard and main rail hub of Los Angeles is a striking structural form. The city's first electrical power station's machine shop and generator factory is today a residence, as well as a monument to architectural ingenuity and contemporary design.

In its eight thousand three hundred square feet of space are spectacular soaring ceilings and glass walls, open steel construction, bent and curved decking on the roof, and two stories of exceptionally dramatic interior spaces. The exterior is a classically proportioned building of stripped concrete and steel; its silhouette is punctuated on the side that faces the rail yard by a huge noise shield of angular corrugated steel panels to protect the residents from the steady clanking and screeching of the railroad. Atop the concrete pavilion is a new steel shell (containing a bedroom annex). Surrounding the building are heavy-duty machines and a salvage yard—this is truly an urban landscape.

The transformation of the 1915 Neo-classical plant was the work of architects Michael Rotondi and Clark Stevens of RoTo Architects in Los Angeles, with plenty of input from the owners, Kathy Reges and Richard Carlson, who had lived there for eight years before undertaking the architectural conversion. (During that time they had retrofitted a former brewery into artists' lofts nearby.) Mr. Carlson is an industrial demolition contractor and recycler—and it is therefore not surprising that many of the most unusual elements of this architectural showplace are serendipitous finds from his salvage yard adjacent to the house. Cast-off trusses, beams, and a staircase were among the recycled features.

"Before" picture shows power station perched above bustling rail yards in Los Angeles.

Conceived as "a series of constructed and volumetric elements," much of the retrofitting took place within the original shell. The emphasis was on improvisation and the unexpected. Familiar elements were assembled in unusual ways, as architect Stevens commented, "to end up with spatial configurations that weren't merely rectilinear and repetitive." Among the most ingenious recycled elements is the swimming pool; the halves of an old gas tank were welded together end-to-end and suspended on steel struts above the newly planted rain-forest garden (which features 40 foot tall bamboo shoots.)

The interior was a challenge, with its 36 foot tall ceilings and vast space. The architects began by reorganizing the space: the ground floor became a gallery for Ms. Reges' growing art collection, while the upstairs is a (somewhat) more intimate living and dining area and a "translucent" kitchen. The master bedroom was added above in a steel shell, while a mezzanine-level guest-room was placed above the living room. A wrap-around walkway joined the units together.

This is an example of adaptive reuse in which the old and the new intersect dramatically; the raw power of industrial steel trusses and contemporary shapes play off the formal lines of the cream cement pavilion, just as the "found" elements are references to a past and have been totally reshaped.

Top: Carlson Reges Residence features original concrete exterior, as well as dramatic steel construction and sound barriers.

Bottom: Soaring interior roof constructions are left visible; overhead armatures suggest mysterious spaces of a Piranesi drawing.

Right: Interior of soaring living room has a striking design with steel armature supporting corrugated metal shell, and contemporary divisions of space; stair behind drywall-encased fireplace leads to guest room and bedroom above.

Below left: Master bedroom with panoramic view of downtown Los Angeles is in a steel shell with corrugated roof and interlocking trusses.

Below right: Gallery for owner's art collection is on the ground floor. Black construction at right is a recycled tank now used as a contemplative space.

THE WAREHOUSE: MICHAEL GRAVES' ELEGANT WORK IN PROGRESS

Chapter Two

Michael Graves, Architect

Over the last 25 years noted post-modern architect Michael Graves has been refining his own adaptive reuse project: The Warehouse, his own residence in Princeton, New Jersey. Today, this once ordinary masonry building is both a showplace and a laboratory for the architect, who tries out many of his ideas here. He was one of the first contemporary architects to retrofit a non-residential building for his own use.

The 1926, approximately 6000 square foot warehouse was built by Italian stonemasons who were working on the construction of Gothic-style buildings at nearby Princeton University. Working in a Tuscan vernacular style, they erected a low L-shaped building, using hollow clay tile, brick and stucco. The interior was divided into many storage cubicles, and there was no electricity, plumbing, or heating.

The warehouse gave Graves an opportunity to explore his interests in formalism, dramatic design, and the relationship between architecture and natural light, and, always, the surrounding landscape. He has involved himself in every aspect of the interior design, as well as the major renovations.

Above: The Warehouse as it looked once renovation had begun.

Left: Architect Michael Graves' house is a former warehouse in Princeton, New Jersey, that he has retrofitted into an elegant residence over a twenty-five year period.

After three decades as an architect, Graves says he has learned that "light is the most important material of all. You must have it, and ideally, be able to control it." The renovation makes use of skylights, French doors, floor-to-ceiling bedroom windows, portholes, classical half-moon windows, and glass brick. The flood of daylight throughout the house "deliberately reinforces the understanding of particular rooms and suggests a continuity between the building and the surrounding natural landscape." The architect uses a selective approach to "energize the interiors with a dramatic sense of the times of day and the times of year."

With its warm pinkish toned masonry exterior behind a welcoming overhang of greenery, The Warehouse today looks much like an elegant Tuscan villa, classical in proportion and detail, but almost cubist in its interplay of planes. The L-shaped structure has two wings: the north wing is now entered through a courtyard, which was once the truck loading dock. As in many of Graves' designs, a series of small spaces forms a progression into the major spaces: today there is a graveled court outlined by a row of trees and terra-cotta pots, a narrow open entrance courtyard, and then a skylighted, formally proportioned rotunda. With the original storage partitions removed, this wing became the living room, dining room and library, with a garden terrace on the first floor, and a master bedroom and study on the second floor. The house also includes a solarium and a guest-room.

The Warehouse is furnished with many items designed by the architect and with his collection of classical antiquities—many of which are displayed in stately niches and shallow wall recesses. It is at The Warehouse that Graves tries out his designs of everything from window shutters to teakettles—more than 400 household items. Thus the architect has converted his warehouse into a thoroughly post-modern combination of classical formalism and updated, welcoming, American domestic design.

Above: The L-shaped exterior of The Warehouse: the pinkish stucco and formal lines of the house recall Tuscany, as well as cubist formalism.

Below: In a typical detail from the exterior, a classical antiquity is carefully placed in a niche above a geometrical form.

Above left: Plantings are part of the overall design. Wisteria vines climb over the columned portico, while espaliered rose bushes ornament another wall.

Above right: Library, with a large collection of architectural books, is itself an architectural statement; high shelving resembles buildings along a city street.

Left: Interior of the living room features deep colors, Biedermeier furniture, chairs designed by the architect, and an antique statue from his collection.

THE NON-TRADITIONAL RENOVATION OF A HUGE NEW ENGLAND HORSE BARN

Chapter Three

Richard C. Tremaglio, Architect

Structural rafters and hand-hewn beams and stairways form a complex pattern overhead and all around. Plants hang from above. Windows bring in natural light from unexpected quarters. The ceiling rises to a stupendous 35- foot height. Niches, nooks, and surprising spaces break up the barn's vast area. Recycled materials mix with the new. A giant, three-story fireplace divides the space and incorporates stairs and hearths. An old tractor seat sits on one landing. In among the aged rafters, new bolts, heating ducts and plumbing pipes are left exposed. This is a non-traditional reworking of a country barn.

When it was purchased, horses still lived in it. Built in the mid-1800s, it had numbered rafters suggesting a traditional New England barn-raising. It needed a new foundation, and about a quarter of the structure had to be rebuilt (with materials from a similar barn). At that point the restored barn was a giant (40 foot x 80 foot) empty space. Architect Richard C. Tremaglio of Boston undertook to convert that emptiness into a livable and interesting space for a family with children.

Working closely with his clients, he came up with a non-traditional approach involving breaking up the huge, "powerful," cathedral-like space in unusual ways, in order to "dam the movement" from one end to the other. Although there are occasional wall divisions forming individual rooms, much of the ground floor is focused on areas of activity surrounding the central fireplace, and a complex series of interacting spaces. Diagonals—both in terms of levels and structural beams—further interrupt the rectangular space of the interior. The dining room, which is built off the east wall, rests on posts and is sunken below the general floor level, rather than walled off. The master bedroom, too, is an addition. A children's room is in the center of the house, while a third bedroom is way up high, like a tree-house; its windows open literally to the tops of the trees.

"Before" photos show the exterior and interior of the original New England horse barn.

A patchwork exterior (reminiscent of 19th century quilts traditional to the same part of the world) is created by a combination of Y-shaped structural columns, additions, eaves, and overhanging roofs, double-hung windows with adjacent acrylic plastic panels, and an exuberant serendipitous quality.

Renovation was achieved as much as possible with recycled materials. Soft pine interior walls (which darken in color each year) contrast with "found" old wood; structural beams and support columns came from a wrecking company and another old barn. Other salvaged materials include oak balustrades and newel posts, a steel bridge, iron railings, sinks, old doors, and steel grates. In contrast to all the interior wood is the huge brick fireplace: two back-to-back units are separated by a staircase (with a "compact, private" space inside with seating, shelves, and growing plants).

Filled with nooks and crannies and unexpected spaces, this barn conversion is an eclectic mix of materials, shapes, textures, and possessions. It is constantly changing. "Intentionally it continues to mellow in character through additions and natural aging. My role as architect was to set that process in motion with a designed framework that allows for changes in habitation. I doubt if it will ever really be finished," comments the architect.

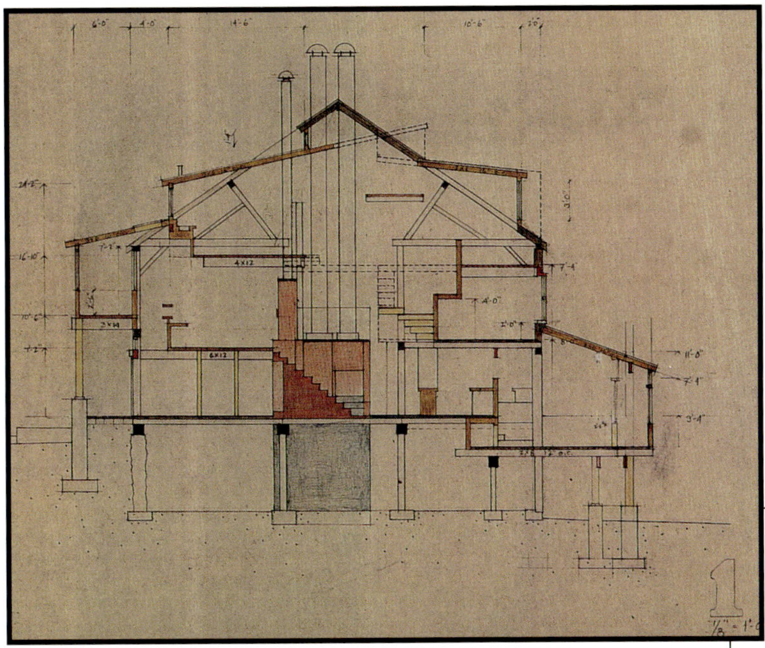

Top: Drawing by the architect shows how the original barn was reconfigured.

Bottom: Interior of the barn with its complex configuration of rafters and beams, windows, and staircases; a central fireplace divides the space.

Right: After renovation: eaves, additions, Y-shaped structural beams, columns, and windows with adjacent acrylic plastic panels to create "patchwork" exterior.

Below: New England barn still in its original farm site, with adjacent sheds. Barn is attached to them in a horseshoe shape.

A STEEPLED MAINE CHURCH TRANSFORMED INTO A WRITER'S HOME

Chapter Four

Maineform Architects and Nellie Blagden

Writer Nellie Blagden came upon the For Sale sign on the front lawn of the church in the town of Richmond while in Maine to read her poetry. She saw the "elegantly crumbling" church with its spacious interior and says it was like "falling in love." She felt that the steepled white New England church represented more than a place to live; her objective in buying it was "to honor the feeling of the space as it originally was." The first time I saw it," she says, "I had a stirring of what it had been and what it might become."

Built in 1857 by Congregationalists, the church had served the congregation until the 1920s; it was used briefly by the Masons, and then by the Nazarenes until the 1950s. The structure was unusually high (eight stories) with four very tall (20 foot) clear glass windows on each side. Though the main body of the building had a traditionally New England style (a blend of Greek Revival and Italianate influences), its 30 foot high steeple was topped by a surprising onion dome. (The Russian-style dome was an 1890 addition after lightning struck the original steeple.) The abandoned church was for sale for only about $15,000; (the congregation, already building a new church, wanted to burn it down).

When Blagden saw it, the dome was precariously balanced 75 feet above the ground, the southern foundation wall had crumbled, the windows were broken, and there was no running water. Blagden consulted Maineform Architecture in Gardiner, Maine, specialists in preservation work. Working closely together, they discovered that 22 feet above the ground floor there was a Victorian scrollwork tin ceiling hidden above a newer addition and an elegant barrel ceiling above the choir loft. They decided together that the best way to retrofit the church for living space was to make essential repairs and then to create a "cocoon" within its vast interior.

Major retrofitting of the structure included removing the dome and belfry before they toppled off (she eventually plans to replace them), shoring up the foundation, removing several recent partitions, and mundane tasks such as insulation and plumbing. The ingenious interior solution to making a cozy living area in such a grand space involved creating a two-story "apartment" opposite the pulpit. The great church hall was retained, with the "cocoon" above. The kitchen is in what was once the vaulted choir loft, and up the oak stairs, in a kind of sealed balcony, there are bedrooms, bath, and a more intimately scaled living room with wood burning stove. This new level bisects two of the great tall arched windows, giving natural light both above and below. The nave area measures 28 x 41 feet (close to the 'golden mean") and provides a beautiful and inviting space that Blagden is beginning to offer for public events.

Blagden has done almost all of the renovation herself. "I was paying for a building, but what I was buying was a lifetime project," she comments about her church. "But when I need guidance, I can go out into the nave and sit or kneel quietly and let the stillness of the place fill me."

Exterior of graceful 1857 Maine church, without its belfry or onion dome.

Above: Balcony and "cocoon-like" apartment creates a cozy niche for the owner in the north end of the vast open space of the church.

Right: Balconied library overlooks nave; gold letters from the past bear the inscription: "I will fill this house with Glory, saith the Lord."

The kitchen is in what was once the choir loft; barrel ceiling and tall arched windows are elegant reminders of the building's past.

Exceptionally tall window height is divided exactly in half so natural light illuminates both the upstairs bedroom and former nave below.

FROM A CALIFORNIA SWIM CLUB TO AN ARCHITECT'S HOME WITH POOL

Chapter Five

Josh Schweitzer, of Schweitzer BIM Inc., Architect

"Before" pictures show the swim club in its heyday; the low structure would become a central pavilion of architect Schweitzer's conversion.

Today, behind a high wall, four detached, one-story structures surrounding a swimming pool make up the most unusual contemporary home of architect Josh Schweitzer, of Schweitzer BIM Inc. in Los Angeles. One of the four houses the family's living area, two are bedroom "pavilions," and the fourth is a studio. The bi-level site is dominated by a huge 60 foot swimming pool—just as the complex was in its days as the ALOHA Swim Club.

Each of the four structures exemplifies Schweitzer's interest in form, color, and texture, and their effects on space (not unlike his noted house design in Joshua Tree, which also combines separate pavilions into a unified whole). With an eye both to creating interesting spaces with simple details and processes, and to working with a modest budget, Schweitzer has converted the low structures of the swim club into a personal oasis in a rundown area of central Los Angeles.

The only rooms in the original flat-roofed clubhouse were a small office, a snack-bar kitchen and the showers. The floor was an uneven cement. Pipes and conduits were all visible. The retrofitting of the swim club began with the knocking down of walls between men's and women's locker rooms and office, and the hosing down of the premises. Then—despite the work ahead—the architect and his wife, a well-known chef, and their son moved in—to one big room with an outdoor shower. "We enjoyed it," says the architect, "because it was like camping out." After that, the step-by-step conversion to living space began. It lasted 12 years.

Now there is more than enough privacy—the property was large, rising from the clubhouse and pool to an empty area with a tennis court—with room for each section to have its own individual space, structural form, and ambiance. And, though each structure does have its own character, the four fit together as an architectural unit as well.

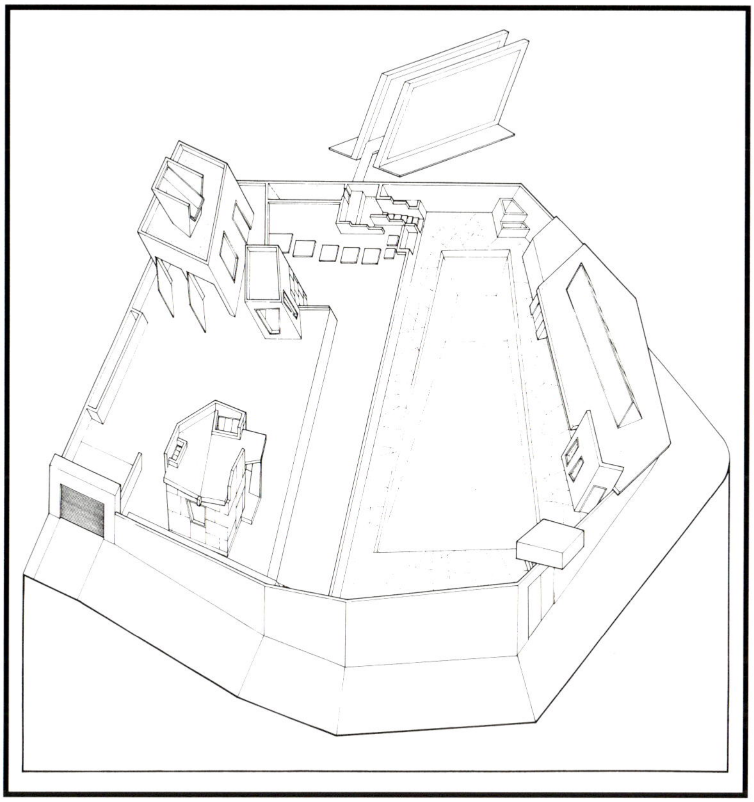

Right: Architect's drawing shows the four separated structures that make up the family's residence.

Below: Living room/ kitchen pavilion now has French doors opening out to the pool.

The living pavilion with an open kitchen area (in the former clubhouse) is characterized by a playful, contemporary, asymmetrical design of free-style angles and shapes and colors—many of them shades of green. Overhead, the soffit ceiling suggests a pool umbrella. Schweitzer lined the walls with broad windows and wide French doors opening to the pool.

The master bedroom is separated from the living pavilion by the pool and a grove of yucca trees. Once the garage, it is a stucco and drywall structure of 520 square feet. The studio, a plywood and drywall building of 275 square feet, has brilliant red and gold walls (the architect describes them as "surprisingly reassuring—beautiful and moody,") and a 16 foot "inverted butterfly ceiling." Recessed lighting and half-walls add to the sense of openness. Their son has his own little house surrounded by banana and bird-of-paradise plantings. It is a smaller version of the master suite, with bright color and abstract, geometric forms within.

The whole family spends lots of time in the pool and in outdoor living in the mild climate of Southern California. The architect, who describes his years of renovating the swim school as "playing," has, in fact, created a home with a distinctly playful feeling.

Top: View from below pictures their son's pavilion.

Bottom: The "living" pavilion of Schweitzer's converted swim club has a soffit ceiling resembling a pool umbrella and contemporary, asymmetrical forms throughout.

FROM A STONE AND WOOD MARYLAND BARN TO A VACATION HOME

Chapter Six

Charles W. Moore and Mark Simon,
of Centerbrook Architects

Originally built circa 1840-1869 on a Maryland farm near the Choptank River, this large barn had distinctive ballast stone walls with hand-hewn wood above, a metal roof, a long shed running along its southern wall, two silos, and a yawning openness within. The owners wanted to use it as a second home for their family of five. They also wanted to retain its character, with the original stone and wood to remain visible from the interior. And the uninsulated barn had to be energy efficient and generally receptive to the sun.

Charles W. Moore and Mark Simon, of Centerbrook Architects in Centerbrook, Connecticut, undertook to convert the historic barn. Their solution to the reuse of the aging structure was an original one. They built new walls outside the old frame walls. The new walls, which were attached to ledger strips and bolted into place, were thoroughly insulated, and covered with a new exterior of rough siding, left to weather. Meanwhile, the original antique wood remained in the interior.

Similarly, the roof rafters were too old and weak to hold up another layer. T-section rafters were added outside the roof, creating space for insulation under a new metal roof.

A major design decision was what to do with the southern wall. The shed running the length of the barn blocked a handsome view of the river and cut off winter sunlight to the interior. The architects chose to strip the siding and most of the roof off the shed, leaving a trellis-like frame, which supports a covering of summer vines, but still provides some shade from the summer sun.

The southern side of the roof became the site for five solar collectors (for hot water), while a conventional oil boiler heats the house. When the house is unoccupied, heat from solar collectors is circulated. Concrete slabs and massive masonry store heat and level temperature changes.

The interior has two living floors and a third floor loft with a catwalk. Three bedrooms, kitchen, and dining room are on the ground floor that opens onto a terrace. A stucco chimney stands the full height of a light well. A generous staircase leads to the second floor, where a bridge across the open space connects the two-story living room, study, and master bedroom.

The strikingly open interior allows spectacular views and sight-lines; the original barn siding, rustic timbers, antique stone, fireplace (featuring metal roofing salvaged from the shed), and barn-like doors are constant reminders of the structure's origins. The architects describe the view from the stair as a "Piranesian fabric of barn timbers."

"Before" photo shows a large barn with tin roof and two silos set in a field near the Choptank River in Maryland.

Adapted barn has new exterior siding (while the old is retained in the interior), trellis-like remains of a shed, and a new roof with solar collectors.

Converted barn's living room, on second floor, has a fireplace using tin salvaged from the shed roof, open beams, spectacular height.

Left: Architects describe the view from above as "a Piranesian fabric of barn timbers." The dining table is visible from a catwalk on the third floor.

Below: Original stonework and barn door in the interior are reminders of the barn's former character.

A BEAUX-ARTS BANK HAS A MODERNIST REWORKING

Chapter Seven

Michael Hauptman, Architect , and Gary Mintz

The 1906 Beaux-Arts bank building on Capitol Hill in Washington, D.C. is a neo-classical gem: it has a palatial quality (it has been likened to the Petit Trianon at Versailles) with its elegant simplicity of design, beautiful proportions, and limestone façade with classical details carved into it. In fact, it was the very image of what a bank should look like—a smallish, very sober and distinguished one—but it had outlived its original purpose and was empty and for sale when seen by the purchaser, Gary Mintz. He needed a good sized space—he collects contemporary art—and the interior of the banking hall had noble proportions, a 20 foot high ceiling, fine natural light from Palladian windows, and about 1600 feet of floor space.

A specialist in adaptive reuse, architect Michael Hauptman of Brawer and Hauptman Architects in Philadelphia, undertook the conversion. But he didn't just renovate the interior; instead, working closely with the new owner, he designed a very contemporary update of a classical space. It is this combination of preserved details and new design that make this one of the most interesting adaptive reuse projects around.

Above: "Before" picture of neo-classical bank in Washington's Capitol Hill.

Right: Exterior of the renovated bank with restored limestone façade and detailing, Palladian windows, and fine proportions intact.

Right: Interior of the bank before retrofitting; high ceilings with egg-and-dart moldings, decaying partitions, and bank vault.

Below: Bold steel fireplace, designed by the architect and owner, can be used by both the living room and study above; it also serves as a strong design element.

Among the preserved features of the exterior were the fine façade with its ornamentation, the wood-framed windows, and the metal cornice. The night-deposit box became the mailbox. In the interior, elegant egg-and-dark moldings were restored, and the great windows and two-story central space retained.

But then came the major changes. They began by removing the suspended ceiling, the carpets and vinyl flooring, and a number of flimsy partitions. The masonry building was reinforced with steel beams. While preserving the spaciousness of the main banking hall—which became the living and dining rooms—they created a new second floor with exposed steel construction; the design breaks up the volume and is "a detail-intensive contrast" to the building's historicism, says the architect.

Downstairs rooms are only suggested, with partial-height walls, while painted metal stairs, bridge, and balcony skirt the central space and lead to the master bedroom and study. The floor of the bridge and study is perforated so the hall (and night lights) below are visible; the height of the second floor brings the original bank ceiling detailing into near view. The two massive bank doors were restored, but now lead to the garden, while a new, less imposing door was installed at the side, so that the entrance is now through the dining room and into the living room. Above the kitchen is a study supported by painted steel open-web joists. A ceramic tile floor now extends through kitchen, dining room, and living room, where it runs right to the hearth of the bold steel fireplace.

The combination of historical detail and proportion with high-tech, functional steel is a bold and appropriate choice for the owner's strong collection of contemporary American art, Breuer chairs, and stainless steel furniture. "I'm very much a modernist," says Mintz.

Above: Open steel bridge and upper level study overlook partially walled kitchen space below; original windows bring natural light to both levels. Traditional detailing complements contemporary structure and art collection.

Right: High-tech stairway to second level study contrasts with nearby Palladian window. Gray ceramic tile covers the floor area.

RECYCLED AND RENEWED: ART AND DESIGN IN AN OKLAHOMA CITY WAREHOUSE

Chapter Eight

Laura Warriner

The transformation of a once gritty neighborhood near the Santa Fe railroad tracks in downtown Oklahoma City is well under way. In fact, artist Laura Warriner had a hard time finding a smallish building for conversion into a live/work studio. She finally discovered a space for which she was able to envisage a dramatic retrofitting.

The 1924 two-story brick building had a typical urban history: it began as an automotive parts supply company. Only six feet from an elevated railroad, it had its own little track running to the building. Then there was a stint housing a print and publishing company. During the 1950s and '60s military uniforms for the Korean and Vietnam Wars were sewn there. (The new owners found needles dropping from the ceiling during renovation.) And then it sat vacant for 27 years. (Shortly before the Warriners bought it, the Oklahoma City bombing—only 2 ½ blocks away—blew out the upstairs windows and damaged the roof, but the elevated railroad next door saved the foundation and helped to prevent further destruction.)

"Before" photos: vacant brick building with boarded-up windows, and state of the interior before renovation.

Warriner wanted a large space for her collage/assemblage work with found objects. The live/work space was to be on the second floor (reached by the original, restored freight elevator). The artist not only wanted to use the space as her own studio; she also planned to open it to visiting and collaborative artists. In addition, she wanted a full kitchen, entertainment and conversation area. At the street level she decided to create a 10,000 foot gallery space for installations, environmental pieces, and performances. There is also a small studio and an office on the first floor.

The artist didn't want to eliminate the original character and integrity of the building, or to make it into "something it isn't instead of appreciating what it is." The exposed beams and wood and brick interior walls were restored and retained. Plumbing and electrical work were contracted out, but Warriner and her husband and friends did much of the rest of the renovation themselves—over a three year period. They redid the red oak hardwood floors, and instead of partitioning the open space, built a low wall of rolling bookshelves to separate the studio from the entertainment area.

In keeping with her interest in recycled materials in her constructions, Warriner created new uses for found objects in her loft as well. Prominent in the loft is the great long dining table, originally the loading dock from the building to the train tracks. The kitchen and work sinks and couches are each recycled pieces of restaurant equipment out of the bone yards of restaurant supply companies. The studio worktables were rescued from deterioration in disassembled piles in a garage. A sliding garage door became the coffee table.

For Warriner, who creates box constructions, and loves the process, or what she calls "the path" of creativity, her expansive loft was simply a much bigger box. "You take an object that has a definition, and you put it in different contexts so it has a different definition. It makes people think of things in a different way."

Above: Exterior of reworked two-story building in downtown Oklahoma City. Exposed beams of the ceiling can be seen in the second floor live/work space.

Right: Rolling bookcase partition (instead of a real wall) retains the open feeling but separates the living area and studio space. Box constructions are by Laura Warriner.

Below: Spectacular new kitchen in the converted space features bare brick walls, open beams, and a recycled sink.

Top: Dining room has open beams, brick walls, view of city; the 24 foot table is a former loading dock.

Bottom: Vast area of the live/work space has no interior partition walls. It features open posts and beams, and red oak hardwood floor, refinished by the artist and her husband.

THE NEW LIFE OF A TRADITIONAL KENTUCKY CARRIAGE HOUSE

Chapter Nine

Clyde Reynolds Carpenter, Architect

The nineteenth century brick carriage house on a city street was a rare and desirable find for architect Clyde Reynolds Carpenter of Lexington, Kentucky. Describing the building as having "considerable charm, good proportions and admirable character," he decided to make it his own residence. The 1897 carriage house is in an area noted for its historic architecture, and the plan was to preserve with "sympathetic renovation" as much as possible of the original structure, while adding on a complimentary wing "that would achieve a sense of continuity."

The building had a large carriage space on the ground floor overlooking a garden, and a hayloft-storage area above. It had recently been used as a garage and had neither plumbing nor electricity. There was an adjacent stable.

The first stage was to remodel the central area. The main space and the stable area became a living room, dining room, kitchen, and service space. Above it, where the hayloft once was, became a library, studio, bedroom and bath, and a balconied living area overlooking the first floor.

"Before" picture shows the white brick carriage house with cupola on a Lexington, Kentucky, street.

"The most dramatic interior alteration occurs in the living room," says the architect, "where the presence of a very unique existing wood-and-metal-tie-rod truss system made possible the easy removal of a portion of the second floor, creating space 27 feet high, and visually relating upper and lower areas."

The dining room and kitchen are separated by a partition made from the old stable doors (which slide with the help of original iron, tulip-shaped wheel supports). Existing brick walls were painted white, while ceiling beams and trusses were left exposed. Three skylights and new glass and wood doors were added. The ground floor (dark-finished hardwood) was raised one foot above the entrance level, both to accommodate ducts and conduits, and to separate and define the living space from the foyer. A new white brick, see-through fireplace also articulates the slate-floored entrance area.

After completing the conversion of the carriage house to living space, the architect began an additional ell. Designed to create a guest wing, to define the garden area, and to provide privacy from neighboring buildings, the addition was conceived as a "three dimensional wall." It visually separates the garden and provides a strong architectural enclosure for decks and brick terraces and a small swimming pool. A sloping, glass-roofed gallery connects the two sections. The floor level of the addition is four feet above that of the carriage house, "permitting several level changes which offset the linear quality of the interior." (The space beneath is used for storage and mechanical systems.) Though the addition's exterior is stucco rather than brick, skylights, which resemble the dormer windows and the cupola of the original building, relate the new to the old. With the most recent renovation, there are nine small skylights.

The architect describes his major goal as "the preservation of the original carriage house while introducing contemporary elements which complement—without imitating—historic design."

Living room has 27 foot ceiling with dramatic iron railing staircase and balconied second story overlooking the space below. A see-through, white brick fireplace separates the entrance foyer from the living room.

Right: Renovated carriage house retains its historic design, but has a hidden addition behind it.

Below: Dining room, made out of what was once the stable, features original wood stable doors and white brick walls.

Below right: Contemporary addition provides privacy for deck areas and garden, as well as guest quarters.

THE RETROFITTING OF A GREEK REVIVAL SCHOOLHOUSE ON LONG ISLAND

Chapter Ten

Elizabeth Thompson, Architect

The owners-to-be spotted this delightful schoolhouse next to the Fire House on the main road of a small town near the tip of the North Fork of Long Island. Its architectural details—from bell tower to original windows, gables, and porches—suggested its era: circa 1900. Not quite as small as the traditional one-room school, but still too antiquated for today's use, the structure had 12 foot ceilings, two classrooms, and a charmingly Victorian interior.

"Before" photo of schoolhouse at its original site next to the Fire House in small Long Island town.

Its Greek Revival exterior was decaying; the old-fashioned interiors were in dire need of repair. Water damage through the leaking bell tower was the major problem; sagging ceilings (despite a previous "jerry-rigged" attempt at shoring them up), and other major repairs were needed. But it had almost all of its interior trim, the original tin ceiling was intact, and even the old slate blackboards and pot-bellied stove remained.

The purchasers moved the schoolhouse about a mile away to a wooded site and hired architect Elizabeth Thompson of New York to convert it into their home. When she began looking over the structure, she found some surprises that figured in the conversion.

An oiled fir floor was discovered under old linoleum tiles. Two wide thick, "pocket doors" were found hidden in the wall between the two classrooms; though one had been destroyed by water damage, the other was still functioning on its track. A mill worker rebuilt the preserved door into two wide leaves; it was placed in the same spot, but in a narrower opening.

The tin ceiling was removed for major structural repairs above it, but it was reinstalled, along with new tin made to match at a workshop in Brooklyn. A new chimney and fireplace were added. A new staircase—to match the floors—was built. The ground floor, divided originally into two classrooms, a girls' and a boys' entrance, and a teacher's office, was converted to a high-ceilinged kitchen/living/dining room, a studio, and a master bedroom and bath. Upstairs, three sleeping lofts and a bath were created.

The exterior was changed very little after repair; a coat of schoolhouse red paint and sharp white trim was applied. It still looks like a country schoolhouse. And although the spacious white interior is now light and airy, the original use of the building as a school remains evident. About 35 linear feet of original slate blackboard still define one wall, and traditional schoolhouse light fixtures and fans were added to the ceiling, but the old, rusty pot-bellied stove was moved outdoors to become a planter.

Right: "Before" interior photo of the schoolhouse shows a pot-bellied stove, slate blackboard, washing station, linoleum floor, sagging tin ceiling, but spacious proportions and large windows.

Below: Finished conversion retained the original Greek Revival architecture; exterior now has a coat of schoolhouse red paint and sharp white trim. It was moved to a wooded location nearby.

Interiors of the renovated schoolhouse have oiled fir floors, white walls, original sliding door, and school room lighting fixtures. Kitchen is tucked into a large living/dining room with the new fireplace that replaced the pot-bellied stove.

A CONVERTED CHICKEN COOP BECOMES A WEEKEND RETREAT

Chapter Eleven

Lawrence Booth, Architect

The Chicken Coop keeps the long, low profile of a mid-western vernacular farm building, but updates it with sophisticated windows and tower.

Set amid prairie grass and reviving oak savannah, this building was once the chicken coop of a farm north of Chicago. Architect Lawrence Booth of Chicago converted the coop into a weekend retreat, as well as adding a viewing tower and, recently, an additional complementary structure for his extended family. Though the long shape of the house certainly brings a chicken coop to mind, most of the original structure's walls and roof were unsalvageable. But the slab and one end of the coop remained (meeting zoning requirements), and setting the shape and dimensions of the structure.

The architect enjoyed combining sophisticated structural features with "traditional crafted mid-western vernacular" farm architecture. The new chicken coop has a number of stylish architectural features that belie its original humble background. Thus the renovated house has the familiar roof line of farm buildings and is built of traditional wood frame and clapboarded cedar siding. But it is stained black, and has an expanse of elegant Pozzi windows with high gloss frames. Above the windows are Chinois panels, to, says the architect "recall a fancy tradition."

The interior is primarily one large single room (measuring 35 by 17 ½ square feet); it has an asymmetrical arched ceiling, a contemporary fireplace, and a full wall of windows that open out to the prairie. An all-glass wall forms the end of the long room and leads to the restored savannah. There are also a bedroom, bath, and small office, as well as a "summer pavilion."

And next to the house is a twenty-four foot tall tower, constructed to enable the owners to "drink wine with a great view of the prairie grasses." (It also provides a vertical contrast with the long low lines of the coop.) The architect recently constructed another house in back of—and perpendicular to—the coop; it is clad in black zinc (zinc, by the way, "never leaks and never needs painting"). Along with the coop and the tower, the new house creates a multi-faceted architectural unit.

Describing the chicken coop and its setting, Booth comments that vernacular construction "connects people to their landscape in diverse, yet profound relationships...This is true, common, democratic architecture."

Natural landscape of Illinois prairie grasses and wildflowers surround the structure.

Long interior is divided into a main room (measuring 35 feet in length), bedroom, and small office.

Above: Arched glass end with an asymmetrical ceiling looks out on the restored savannah. Pozzi windows open out on one side; contemporary fireplace is opposite.

Below: View of coop, with additional "zinc house" at right angles to it, shows the setting and the 16 foot tower erected to enable the owners to "drink wine with a great view of the prairie grasses."

A GARAGE ON BEACON HILL GETS A MUSICAL AND ARCHITECTURAL REWORKING

Chapter Twelve

Graham Gund, Architect

Among the orderly town houses in the well known and gracious Boston neighborhood of Beacon Hill (where living space is at a premium) are many small private garages. Sometimes surrounded by tiny private gardens, such garages are tempting targets for retrofitting and conversion.

Architect Graham Gund of Boston undertook such a transformation: the Deutsch Residence is a former garage with its own courtyard. Attached to the neighboring house by a high brick garden wall, the garage formed a corner; two large, gently arched doors opened to the residential street. The structure was to be converted into a living and working space for two musicians.

While retaining the distinctive brick, one-story garage, Gund added another floor with a stucco exterior atop the building. The stucco addition had Tudor-style decorative cladding and a steeply pitched roof, giving the tiny structure its own architectural identity, as well as doubling the living space.

On the original garage floor there are now music studios for the owners and a one-story master bedroom. A two-story stair hall connects the ground floor with the new second floor, where an "open plan arranges living space around a central hearth, which is the symbolic heart of the home."

But this is not simply an open, loft-like room. The space is divided and refined with an eye to angles and spatial relationships. In contrast to the strongly "grounded" contemporary fireplace is a curved study loft that rests on light columns and seems to float above the room. High, peaked eaves create an additional sense of space.

At the very top of the roof the architect paid tribute to the "artistic energy and exploration within" with a sculpted musical staff (including a G clef and notes). In fact, he sees the entire structure as an exploration of space and "the poetry that is the fruit of music and architecture."

"Before" snapshot shows a Beacon Street neighborhood with the one-story garage before conversion.

Right: Drawing showing how the architect used space.

Below: Exterior view shows what once was a garage transformed into a two-story house and garden.

Interior of the second story shows interesting spatial relationships with eaves and curved "floating" study loft, which creates "visual tension" with the firmly grounded fireplace unit (seen here from the back). The kitchen is on the other side of the loft division.

AN ART DECO MISSISSIPPI
CAFÉ GETS A REWORKING

Chapter Thirteen

Samuel H. Kaye, Architect, Gayle Rutledge and Marc Merciera

"Before" photos show front façade and back of the old café building. Interior snapshots reveal sad condition of walls, floors, ceilings and wiring due to water damage from a leaky roof and a century of neglect.

At the end of a city block in the Historic District of New Albany, Mississippi, sat an old street café in dire need of renovation. Its interesting architectural details caught the eye of decorator Gayle Rutledge and her son Marc Mercier, offering a challenge to these imaginative buyers. Originally constructed around the turn of the century, the structure was first renovated in an "Art Deco" style some fifty years ago. The small café had also had a past as a pizza shop and a pawnshop. A two-story stucco and pre-cast concrete structure, it had an overhang at the entrance to the café and some very distressed empty spaces on both the first and second floors.

The buyers chose architect Samuel H. Kaye of Columbus, Mississippi, to transform the building. While not altering the total floor space of 3510 square feet or the basic exterior architecture of the building on the historic block, he had lots to do within. The decision was made to use the ground floor as an office space, and to create living space on the second floor.

Though dirty and cracked, the exterior of the building was notable for its design of white stucco with unusual narrow vertical black glass inserts and glass tiling, a nod to Art Deco elegance. The owners renovated the exterior; a trim new coat of stucco replaced the decayed façade, and a row of tiny lights was added to border the overhang and to provide a touch of night time glamour. One wall of the building was broken through to provide an upstairs exit; an exterior stair and balcony of wrought iron added a decorative note. A new roof was a necessity.

The condition of the interior walls (see "before" photo) was deplorable; years of neglect had taken a toll. Wiring was exposed and dangerous. Water damage was everywhere. There was no kitchen.

But there were surprises too. When indoor siding atop a layer of plaster was removed, beautiful old brick walls (common to this part of Mississippi) were discovered. While stripping layers of decayed wallpaper and other materials from the ceiling upstairs, the owners were delighted to find original fine wood from the turn of the century, as well as a giant skylight that had been thoroughly plastered over. Elegant archways, good flooring, and the original coal fireplace (once the only heat for the building) were restored. A new kitchen was added beneath the skylight. The Rutledge Mercier Building now provides attractive living and office room with a touch of elegance and Southern charm.

"After" exterior photo shows a redone "art deco" façade with its narrow vertical black glass inserts, lighted overhang, and glass tile.

Right: The original coal fireplace and damaged ceiling were updated; a fine flooring was found under the debris.

Below: When cheap siding and layers of plaster were removed, the downstairs walls were found to be brick with attractive arches.

Bottom: An unexpected skylight, found when layers of wall paper were removed, illuminates the new kitchen.

THE SURPRISING TRANSFORMATION OF A CONNECTICUT CISTERN

Chapter Fourteen

Peter de Bretteville, Architect

The abandoned cistern had what the architect, Peter de Bretteville of New Haven, calls "real presence." Built in 1907 to store water, it sits imposingly on a ridge above New Haven and the Sound—a massive, circular, concrete structure with its own tower. For the de Brettevilles, who were seeking a loft to remodel, it presented an opportunity to convert a solid and most unusual structure into a spacious home and workplace for themselves—as well as a dynamic architectural statement of preservation and renewal.

The original cistern was 54 feet in diameter and 28 feet high—a massive, yet workable size. When they bought the derelict building it was windowless. Its walls, which were 18 inches thick, were crumbling. It had a dramatically deep cornice, its own 66 inch wide hexagonal stair tower up to the roof, and four distinctive supporting columns within, Needless to say, it was very dark and damp within.

"Before" picture shows an abandoned cistern on a ridge above New Haven.

The conversion of the cistern combines preservation of the original distinctive shape and proportions with a light and airy solution. The circular structure and tower were retained but the exterior was renewed with insulation and fresh stucco of a Tuscan yellow-orange tone, while the interior walls were sand-blasted and treated with lime wash. Openings were cut into the thick walls and roof and industrial-style, aluminum-framed windows added, including four circular skylights and a series of double-height French doors and clerestory windows. A columned, trellised terrace (using salvaged material from the reconstruction) was built to overlook the city and water below; a canvas canopy is strung from the top of the vine-covered columns in summer months. A terrace was created on the roof, as well.

In the interior, the architect divided the space into two stories; the first floor has living and working areas, while the bedrooms are above. On the ground floor the interior design was created around the couple's vast library; they own 500 linear feet of books. The four original supporting columns created a 16 foot area which became a dramatic two-story high library square contrasting with the circular cistern walls around it. All of the downstairs rooms—including their two pie wedge-shaped offices, the east-facing kitchen/dining area, and the two-story high living room have the distinctive round outer walls and are flooded with natural light. The sand-blasted walls have a pale greenish-white tone, while the original ceiling is bare concrete; both materials contrast with new interior construction dominated by heavy timber framing.

A steel staircase leads to the partial second floor. A glass and steel bridge or catwalk—lighted by skylights—connects the bedrooms and guest suite above and looks down on the floor below. Upper story windows, skylights, and glass brick bring natural light into bedrooms and baths.

"In the past," comments the architect, "buildings for public works were built as temples." While the interior is sparklingly contemporary, from the outside the old cistern—where once 365,000 gallons of water were stored—indeed resembles an antique Italian temple or villa: elegant, vine-covered, warm-toned, with its own dignified geometric presence.

Below & Opposite top: Exterior views today: the cylindrical structure still has the original decorative cornice and hexagonal tower; updated structure has a new trellised patio, new windows, double-height entry, and newly stuccoed exterior in warm yellow-orange tone.

Above: The master bedroom is connected by a skylighted glass and steel catwalk to a guest suite; bridge overlooks library.

Left: Interior has a two-story living room with distinctive circular walls and a dramatic central library defined by original supporting columns.

A WISCONSIN DAIRY BARN CONVERTED WITH RECYCLED MATERIALS

Chapter Fifteen

Tom and Bobbie Haig

Over the past twenty-five years Tom and Bobbie Haig have been renovating a massive Wisconsin dairy barn with an attached tiled silo. This is not just an ordinary, run-of-the-mill barn; it is 100 by 32 feet in size. It was on an 80 acre farm (with a run-down old farmhouse as well) when the Haigs bought it. They moved into the farmhouse until the "front end" of the barn was habitable, and since then, renovating section by section, have turned the barn into quite an awe-inspiring place. It now has two levels of living space, a ground level workshop, laundry, furnace room, garage, and a circular staircase to the second floor that rises through the original silo.

Each step of the way the Haigs photographed the progress they were making. The pictures tell a story of ingenuity, recycling, and hard work, as well as a taste for spacious, high-ceilinged barn architecture. Tom Haig himself did all of the design work, most of the labor (including wiring, plumbing, tile installation, fireplace construction, and even the building of the spiral staircase). Recycled materials were collected by both Haigs; Bobbie Haig found old bricks people were throwing away and scrubbed them down for the new fireplace, and her husband was alerted to large pieces of marble being discarded by a hospital. (The marble later became part of the kitchen and bathroom and fireplace wall.) Segments of an old bowling alley became tables and a workbench. "We always have people looking out for stuff for us," said Mrs. Haig. "It's glorious."

Above: "Before" picture shows a large wood cow barn and tile silo (with damaged top) as it looked when the Haigs purchased the farm.

Below: Transformed barn has a greenhouse, deck, garden, repaired silo, and a huge amount of interior space.

The Haigs photographed stages of progress in retrofitting the barn: 2 x 12 joists were used to push the stone wall back into place; the silo's damaged top was replaced; the west end of barn got a temporary plywood wall so a windows wall could be built behind it; the interior was insulated and the barn received new inside framing and room divisions.

Major retrofitting included repairing the stone foundation, which no longer supported the barn; 2 x 12 joists were used to push it back over the length of the summer. Haig also put up a temporary plywood wall before building a permanent window wall behind it. The silo's top had to be replaced. Insulation, new interior framing, and room divisions were essential additions.

The conversion of the barn interior began with the building of a kitchen, breakfast and dining rooms, two bedrooms, a den, and a bath in one end of the barn. After that, a fireplace and chimney went into the great two-story tall living room. In addition, the first floor now has a master bedroom (with a Japanese motif), an exercise room, and even a greenhouse. An expansive "martini" deck adds yet more space.

The second floor areas at each end are joined by a balcony that looks down into the massive living room (the railing was recycled from the communion railing discarded and auctioned by a nearby Catholic church). Each end of the floor has two bedrooms and a bath. There is an additional library and den on this floor too.

The spiral staircase connects the two levels. It is suspended from the ceiling of the silo; at its center is a plastic tube that distributes heat and light. And at the very top there is a little "hideaway" with a great view. Clearly, the Haigs like a lot of space around them. And certainly the use of recycled materials seems a fitting and imaginative response to the recycling of such vernacular architecture as this typical dairy barn in Wisconsin.

Below left: Two-story high main living room features a balcony with railing salvaged from a church, open beams and lots of natural light.

Below right: The silo now houses a spiral staircase; in the staircase's center is a plastic tube that distributes heat and light.

THE TRANSFORMATION OF A CARRIAGE HOUSE AND STABLE IN VIRGINIA HUNT COUNTRY

Chapter Sixteen

Mark A. Kohler, Architect

"Before" photos capture the old carriage house from front and back. There are two wings of stables around a courtyard; typical stall entry is underneath an overhanging roof.

Goodstone Farm's turn-of-the-century carriage house and stable was a large, once elegant building in ruins when purchased as part of a 265 acre Virginia farm in the foothills of the Blue Ridge Mountains. The historic structure accommodated three carriages, ten stables, two additional stalls, and a small apartment. Its two wings of stable created a natural courtyard. The building was in disrepair: many of the exterior walls were rotted and termite-infested, it needed new windows, a new roof, and new floors. But its fine proportions and architectural detail made it well worth adaptive reuse.

The farm had had a long and interesting history; it was purchased in 1768 by "Jamie the Scot" Leith, provided food and comfort to the Continental Army during the Revolutionary War, and sent off three Leiths to fight for the South in the Civil War. During the 1900s, the farm became an award-winning dairy. In the last fifty years it had been the fox-hunting headquarters of its region. But when purchased recently, the original mansion was a vine-covered ruin, and only the carriage house and stables were salvageable.

The new owners wanted to turn the carriage house into a "livable, comfortable residence." Architect Mark A. Kohler, of Karl E. Kohler Architects in Burke, Virginia, undertook the conversion. He made an effort to preserve the overall architecture, including the signature Palladian window design, and the two long wings with covered cobblestone walkways where the stables had been. The center

Retrofitted carriage house has the same architectural design but the three carriage doors are converted to Palladian style glass doors under original half round windows. Stable wings are now bedrooms with stall doors opening to a cobblestone walkway.

courtyard was retained. All of the ten antique wood stable doors were left intact, but the upper halves were fitted with casement windows. The three front doors at the center were replaced with glass doors, directly below the existing half-round windows. In the back of the building, closed walls were opened up with more glass and the elevation raised.

Six bedrooms were placed in the original stable wings. The carriage area in the center of the structure became a spacious living room for entertaining, or "great room." A big stone fireplace was added. Both the living room and nearby large country kitchen open out behind the house to a new deck resting on stone columns. The master bedroom and bath suite is upstairs.

The retrofitted interior matched the original design whenever possible: old woodwork of horizontal tongue and groove was used to replace the inside walls, plumbing fixtures were chosen that were consistent with the time period of the building, dark stain colors from the original were utilized throughout, and all interior doors and hardware were duplicated.

Having converted the building, the owners opened it to visitors; it became The Goodstone Inn. Guests, who are staying in what were once the stable stalls, will no doubt find horses nearby, for the site is once again a center of activity in this fox-hunting region.

The back of the structure has matching glass doors overlooking the countryside and a new raised deck resting on stone pillars.

Interior of the "great room," where carriages once were kept, features lots of glass, a large stone fireplace, distressed heart of pine flooring, and horizontal tongue and groove walls and ceiling.

THE UNUSUAL UNDERGROUND CONVERSION OF A MISSILE SILO

Chapter Seventeen

Edward and Dianna Peden

Above: The entrance to the underground home leads into a hillside, through a 47 ton garage door leading to the entrance tunnel.

Below: A 120 foot long rounded tunnel leads to the Peden's home in a former missile base.

The owners, Edward and Dianna Peden of Dover, Kansas, call their home underground a "Twentieth Century Castle." Peden likens the "hardened defense-works of our century" to the medieval castles of Europe. He not only lives in what was once a 1950s missile silo, but he sells other such sites to people who want to live in them as well. These first generation missile sites, he says, are "unique, heavily reinforced historic structures; designed to withstand nuclear attack, they bring new meaning to the word 'shelter'."

The United States built quite a few of them during the Cold War (there are others in Nebraska, Texas, Kansas, and New York); these Atlas and Titan sites were decommissioned after 1965, to conform to the S.A.L.T. Treaty. Worldwide publicity followed the offering of the sites to private individuals, and the Pedens were among the early converts to underground living, retrofitting their site for family life. Their silo, constructed before the Cuban Missile Crisis, is one of nine such sites in Kansas.

Most of the sites are in somewhat "rough" condition after 30 years of neglect, but "with some clean up and reconstruction inside, their grandeur is restored," according to a Peden brochure. The "Atlas E" sites are the most coveted for retrofitting, with 15,000 square feet of floor space in two main underground buildings connected with a 120 foot long tunnel. There is a drive-in access through a 47- ton garage door. Walls and ceilings are 18 inch thick, heavily reinforced concrete. The buildings are covered by 3 to 5 feet of earth. Set in some 33 acres, Peden says "they offer the ultimate in engineering and design."

The Pedens live in what was once the massively large home of an intercontinental nuclear missile, at the end of a great tunnel. Reconstruction involved pumping out 8 ½ feet of rainwater, and caulking the roof (a 400 ton concrete and steel retractable ceiling panel). Their house is now lighted by a giant 128 square foot skylight, for there are, of course, no windows. But they also built two towers to provide them with views of the out-of-doors.

They have built a two-level home with a dining area and kitchen, a living room, and four bed-

Overview of the Kansas landscape shows one of Peden's light aircraft above the hidden missile base with its underground entrance and ground level roof.

rooms within the 18,000 square feet of space in their silo—using only 2300 square feet of it. They have considered turning the 6 foot wide trench that was once a cooling bath for Atlas engines into an indoor wading pool. In the adjoining missile bay Peden builds ultra-light planes (the property has its own airstrip too.) Their atrium is greenery-filled and light due to windows in the tower. Atop the home is their patio, with grill. Underground living means, of course, that the house is cool in summer, and in need of only a wood burning stove in winter (though electric costs are high).

In this example of adaptive reuse, it is easy to remember the original use of the building; aside from the underground locale, the tunnels, and the vastness of the site, the Pedens have chosen to retain certain aspects of Cold War decor including "Fallout Shelter" signs and warning signs of danger. But the interior of the living space is remarkably normal looking!

Above: The owners constructed a giant skylight to bring natural light to their underground home; in the center, an atrium is light and filled with greenery.

Below left & right: The kitchen and dining area, though they have no conventional windows, are comfortable and lighted from above.

Chapter Eighteen

AN OLD STONE SMOKEHOUSE AND ITS "INTERNAL TRANSPLANT"

Steve Badanes and John Ringel of Jersey Devil

When they purchased a house and a piece of land near Princeton, New Jersey, the new owners found a small, very old stone structure with a dirt floor in the back of the property. Apparently built some 125 years earlier and used as a smokehouse (or possibly an icehouse), the building had stone walls, one story, and the dirt floor that depressed four feet below the threshold. The gabled roof, however, rose to 14 feet, where a series of tie-beams held it together.

Steve Badanes and John Ringel of Jersey Devil in Stockton, New Jersey, undertook to convert the structure into a study and occasional guest house. But they needed to add both a second story and a new floor. The major problem was how to add a second level while maintaining the rafter bracing.

Their solution was ingenious. Into what Ringel described as a "startling void," the architects created a "tensile" structure that appears to float in the air. In fact, it was a prefabricated form, "a handcrafted, wood sculptural item suspended in a neutral, sheet rocked, Monopoly board house-shape volume," as described

Exterior of an old stone structure, once used as smoke house or ice house, now a guest house and study.

Left: The kitchen includes a counter and stools recycled from an old diner.

Below: Architect's sectional concept drawing shows the combination rafter tie and hanger for the second floor, enabling the volume to accommodate two stories

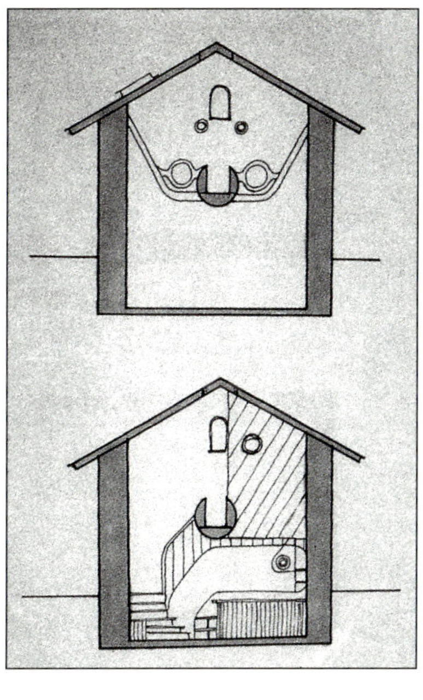

by Badanes. This "hanging cradle" had sloping sides; the end of each "arm" was connected to a rafter, thus forming a kind of loft. It connected to a "bridge" and to a handcrafted fir stairway.

Skylights bring in natural daylight directly over the loft "arms" (contributing to the illusion of floating), and new arched windows were added at each end of the upper level. Contrasting with the airy and light second floor, however, is the dark and cozy ground floor, where a recycled diner counter with stools serves as the kitchen counter, and where there is a freestanding metal fireplace and a curved plexi-glass window overlooking a new swimming pool.

Described as an "internal transplant," this conversion of an old building is both unusual and ingenious. It also has a certain humorous element: the staircase has "hiding holes," and the circular loft railings resemble a large pair of glasses surrounding the "nose" of the loft's end.

Above: The ingenious design for a sleeping loft has curved supports; new skylights and end windows create a light, airy environment.

Below: The suspended hanging "cradle" is a handcrafted wood sculptural form.

AN OLD TOWN ALEXANDRIA FIREHOUSE RECLAIMED

Chapter Nineteen

Joanne Goldfarb, Architect

The volunteer fire station was a narrow 1855 Greek Revival building called the Relief Fire House, on one of Old Town Alexandria, Virginia's, well-preserved streets. It was a typical small town firehouse of its day: made of a warm red brick, it had an open first floor space for the horse-drawn pumper and a meeting hall (or fire marshal's office) above. The whole structure was only about 18 feet wide and 50 feet deep. (Once motorized fire engines arrived, its narrow dimensions made it obsolete; it was replaced by a larger station next door in 1915.)

But it definitely had the kind of character that appealed to architect Joanne Goldfarb of Alexandria, who bought it at an auction and converted it to living space for herself and her family. "It was irresistible," commented the architect. "What I liked about buying a firehouse, was that it has a history, because it comes with logbooks and records." And it also had "some extremely tasty elements to deal with." Among them were a 16- foot high second floor ceiling, large southern-exposure windows, and fine exterior wood detailing typical of its period.

Goldfarb tried to retrofit the firehouse as much as possible without losing original interior and exterior features. She depended on historical photographs to recondition the exterior. Sometimes what she couldn't save she rebuilt, patterned on the original. "I was very intent on letting old things stay as they were," says Goldfarb. Among the preserved features were the original over-scaled baseboards, pine floors (which were sanded smooth), and some stripped brick walls.

Exterior of a trimly refurbished Greek Revival firehouse shows a new vestibule and replicated garage doors, as well as a restored second floor balcony with holiday bunting.

Right: "Before" photo of the interior shows a high ceiling, open space, and the neglected condition of the firehouse.

Below: Barrel-vaulted skylight measuring 18 x 6 feet extends across the width of the building, providing light to all three floors.

In order to preserve the historic exterior of this landmark building she consulted antique photos of the balconied original; she replicated firehouse garage doors to flank the recess containing the front door, and restored a black iron second floor balcony bearing new cast aluminum letters identifying the "Relief Fire Company." On holidays she hung the balcony with old-fashioned red, white, and blue bunting. She kept the building's large front windows.

The ground floor now contains the dining room, with a wood and glass wall separating it from the kitchen. A cast iron spiral staircase rises to the second and third floors. A new mezzanine level inserted on steel beams enabled Goldfarb to stack three bedroom and bath levels in the back of the house. The loft-like master bedroom is at the top.

Most of the second floor is open. Here, the spacious, light living room has French doors and a 16 foot ceiling rising to a barrel-vaulted skylight measuring 6 x 18 feet. There is also a library behind the living room, under the new mezzanine floor.

While in the interior of the house the overall impression is a contemporary setting of glass and architectural space, Goldfarb included distinct reminders of the structure's past: an old fire tower wooden ladder (a gift from the next door firehouse) leans against the living room wall, a 19th century industrial pulley hook hangs 3 stories from a skylight beam, a vertical pole (both symbolic and structural) rises through the center of the stair, and the original metal tracks that guided fire engines' wheels are still embedded in the floors of the kitchen and dining room.

Above: The living room is on the second floor; French doors open out on a restored balcony. A spiral stair with pole and ladder suggest the firehouse origins within a white, spacious, contemporary interior.

Left: First floor dining room features a bare brick wall, new glass and wood wall dividing it from the kitchen, a spiral staircase around a symbolic and structural pole, and a dramatic industrial pulley (used for hauling furniture up and down).

Chapter Twenty

OPEN, LIGHT, AND COLORFUL: THE RENOVATION OF A LONDON POST OFFICE

Gunnar Orefelt, Architect

The vast spaces of an abandoned post office in London became the perfect setting for an artist's home and studio in this sophisticated adaptation. Open interior space is at a premium in London, as in so many cities, and when the artist set out to search for live/work/collection display—and entertainment—space, she was delighted to find the old Victorian P.O. available. Once a sorting station for the Royal Mail, the building's interior was now vast open space. It had high sky-lighted ceilings with a spider web of struts, a floor space of 1700 square feet on the ground level, another 1250 square feet on the floor above, and 300 square feet in the basement. The brick exterior was punctuated with attractive tall arch-topped windows.

Architect Gunnar Orefelt of London undertook the conversion. His objectives were to create an environment that would be free- flowing spatially, lively in design, and inviting both as a work space and welcoming home. The artist needed a large studio to work on her oversized canvases (and a 40 foot storage area), plenty of room to display her collection of contemporary art and "vintage clothing and Mc Donalds giveaways," a space for entertaining (including a pool table), as well as a library, bedroom, guest room, living room, dining room, kitchen, and even a courtyard and rooftop terrace.

Victorian exterior of an old sorting station of the London Post Office.

Original Post Office interior with an overhead spider web of struts, lighting and skylights.

Orefelt chose to create the most minimal of divisions in the vast space. He used half walls and plate glass to retain the natural light and airy atmosphere, and he added more skylights and backlighting—hidden florescent lighting behind panels and cupboards. He retained the geometrically interesting overhead struts.

Color is everywhere. Instead of the cold white minimalism of many loft designs, the artist and architect decided on a more colorful and inviting ambiance. There are new terrazzo floors, bright colored partitions, brilliant yellow kitchen cupboards, peeling red furniture recycled from an old French sanitarium, and even a red and blue '60s "Pop-art" target table. (One bathroom with vintage fixtures is turquoise, another jonquil yellow.) Even the owner's clothes—in a glass storage area—are hung by color.

The lower level includes the studio, art storage, library, and bedrooms. A free- standing stairway leads to the main floor where the living areas flow openly from one to the next, except for a separate dining room. A rooftop terrace (of 300 square feet) completes the renovation.

The owner, who is still redecorating parts of the interior, describes it as "such a welcoming house." Besides providing the artist an exceptionally light place to paint, the vast open space lends itself to entertaining—but also to quiet moments. "It's fantastic for a party," she comments. "There are speakers in every room and the whole house is made for dancing. At other times I find visitors just clam down when they're here. It's a house with a very good vibe."

Right: Space is divided into two levels, featuring low partitions, brilliant color, and original post office windows.

Below: Living room today is exceptionally light and spacious, divided from other sections of the floor by low partitions.

Dramatic staircase leads to circa 1965 table painted like a target.

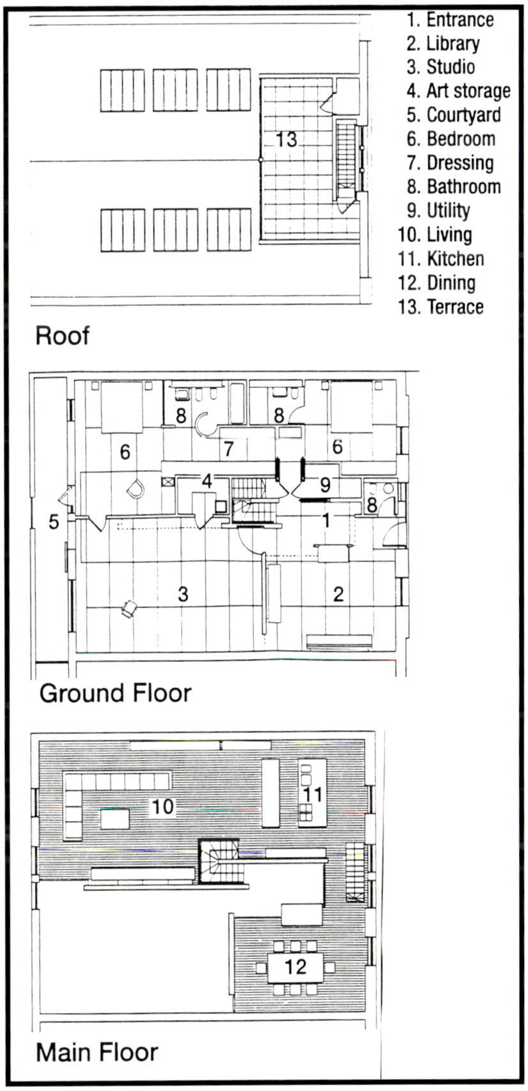

Drawing shows how the vast interior space was divided.

THE SAVING AND RENEWAL OF THE ROSE ISLAND LIGHTHOUSE

Chapter Twenty-One

Charlotte Johnson and the Rose Island Lighthouse Foundation

The American lighthouse has a romantic image, beloved on both coasts, as well as along the rivers and lakes of the nation. Its often lonely setting and welcoming beacon, and the isolated, self-sufficient life of the lighthouse keeper, have a long historic tradition. What better way to recreate that quintessential "alone with nature" experience than to live in a converted lighthouse?

The maritime architecture of the nation's lighthouses varies somewhat, but is often a familiar blend of small, white wood-shingled New England farmhouse with tall, chunky tower—just the way we imagine it. More and more of these old-fashioned buildings are being decommissioned as technology makes them obsolete. A prime example is the Rose Island Lighthouse, which sits on a 16-acre island in Narragansett Bay off Newport, Rhode Island.

Rose Island Lighthouse, circa 1900: it had a colored slate mansard roof and a fixed red beacon. It sits atop a former fort. The A-frame tower was built in 1885 for a fog bell.

Built on the remains of an 18th century fort, and first lighted in 1870 as a link in a chain of lighthouses along the New England coast, it was active for over a century. When a bridge was built nearby, the lighthouse was abandoned by the Coast Guard; vandalism and exposure almost destroyed it before a group of public spirited citizens—under the leadership of Charlotte Johnson—went to work to preserve it (a giant condo complex was planned for the island instead.) "It just seemed so destructive of that environment," commented Johnson, who choreographed a 100 person white-sheet encircling of the island in protest.

The nonprofit Rose Island Lighthouse Foundation was created and spent some eight years raising money, and volunteering hours and materials to retrofit the building—a giant task. They succeeded. They even relighted the beacon, today powered by wind-driven generators. Now on the National Registry of Historic Places, the lighthouse, which is only reached by boat, has a new keeper: Ms. Johnson. It is open to the public as a museum and educational center, and welcomes occasional guests who want to work on the place. Only ten people at one time can visit the island during bird-nesting season.

At the start of reconstruction the lighthouse had a leaking roof, deteriorating plaster, fifteen years' of bird droppings inside, a rotting cistern, and decaying exterior walls. The interior was equally in need of repair.

A new slate covering for the distinctive mansard roof was one of the first improvements. New exterior siding covered by clapboard followed. The old-fashioned trim was restored and painted. The interior was cleaned and fungicide applied. Storm windows and insulation were added. Wintertime heat comes from passive solar windows plus a radiant under-floor oil heating system (the keeper's apartment also has a wood burning stove.) The entire floor and joists had to be replaced and the interior walls were covered with sheet-rock and new plaster. The new wind-powered generator is the only electrical source; it is augmented by candle power when necessary. A list of repairs still needed keeps visitors busy—but they also get the nostalgic treat of living for a few days in a lighthouse.

Left: Reconditioned lighthouse, powered by a wind-driven generator (to the right of the flagpole), once again sits proudly on its island in Narragansett Bay.

Lower left: The lighthouse before restoration; though the roof leaked, the trim had been removed, the clapboard covered over with shingles, and birds were living inside, it still had a lonely beauty.

Lower right: Interior before renovation, with falling sheet-rock, plaster and bird droppings. All of the original plaster walls had been covered over. Five layers of linoleum covered the kitchen floor.

Above & right: Interior today: kitchen is now stocked with antique implements and an old coal stove, though there is also a propane burner for cooking and heating water. Walls and stair have been newly redone.

Far right: Doghouse replicates both architecture and water-gathering system of the lighthouse: rain washes off the roof, into the gutter and downspout, and into the dog's dish.

CAPTURING THE CHARM OF AN 1860 COUNTRY TRAIN STATION

Chapter Twenty-Two

Lois Bradford, and Joan and Dennis Sawyer

Among the most beloved of American buildings are its country railroad stations, often nostalgically seen as charming remnants of a romantic past when trains whistle-stopped in rural hamlets across the nation. Every town had its little station, frequently designed in Victorian fashion, with dark wood siding, overhanging roof, and even gingerbread trim. Alas, most of these stations were obsolete some years ago, but here and there train buffs, Victorian architecture enthusiasts, and intrepid do-it-yourselfers have given them a new life.

This station, in Sheffield, Massachusetts, had given 110 years of service to the New Haven and Hartford line through the Berkshires, but then passenger service was discontinued. Though offered for sale for just one dollar a few years ago, the buyer had to move it from its trackside (freight trains were still using the line). The move cost $9000 (for a two-mile trip), but the owner, Lois Bradford, had a true piece of history to retrofit into a spacious and unusual home. She set about to do so while preserving as much as she could of the original structure.

Lost in the move was the original gingerbread ornamented roof –which had to be removed so that the structure could travel under the trees. Also now missing are the two original potbellied stoves that warmed the waiting passengers. A new foundation was needed, and the building now no longer sits at track level.

But the rest is architecturally intact—one recognizes it as a rural depot immediately. Retained are the original dark wood façade, great high-pitched roof line, the 15 foot tall ceilings, and the grand open space of the interior. The rounded ticket windows are now part of the kitchen. The bow windows along the side (from which an arriving train could be seen in the distance by indoor viewers) are still there. Small octagonal shaped rooms (once the mens' and women's bathrooms) have been retained. The stationmaster's office is now a cozy den. Dark wood molding and arched windows have been restored.

The original open space has been divided by a few new interior walls and a staircase to a major new addition: a balcony with bedroom and bath, and a great overlook of the original waiting room. In one end of the structure another bedroom has been created with built-in dark wood resembling an old-fashioned sleeping berth in a Pullman car. Two fireplaces (and updated heat) now take the place of the old wood stoves; there are two new chimneys as well.

The present owners, Joan and Dennis Sawyer, are enthusiastic about the depot's past. (They run the Depot Guest House next door and offer tours of their station house home.) And, not unexpectedly, along one side of the vast living room—where passengers once gathered—is an impressive array of model trains and tracks ready to set in motion.

Sheffield (Mass.) station in its working days. Ornamental gingerbread trim and high pitched roof were typical of its Victorian period.

Right: Octagonal window allowed passengers keeping warm in the waiting room to look out and see trains coming in the distance.

Below: Station today has been moved nearby (away from tracks still in use); though gingerbread is gone, octagonal ticket window and body of the structure are intact.

Left: Interior of great waiting room now includes living and dining rooms; original station master's office is now a den (left door), while kitchen (right) is in the octagonal former ticket office. A quilt-draped balcony overhead looks down on the great room.

Below left: Living room, with original high ceiling, dark moldings, and tall windows, accommodates owners' collection of model trains.

Below right: Built-in wooden sleeping alcove resembles a Pullman berth.

MORE SINGLE FAMILY HOME CONVERSIONS

The following seven projects suggest ingenious solutions to adaptive reuse, some still in progress. In each of them the original building has been converted for living from its original use, and in each, the retrofitting has been suggested by the architecture and history of the building. Several of the projects are small scaled, but have resulted in cozy residences that have an aura that would be impossible to recreate from scratch: a rural one-room schoolhouse, a small country barn, a 19th century village church. In two projects, the architects have surprised us with unusual make-overs within unexpected exteriors: in one a large, complete antique barn is hidden within a new exterior, and in another an ordinary-looking machine shop on a city street has a brilliant new interior. Two projects—a stone garage and an antique canal house—provide homes and studios or galleries for working artists.

A ONE-ROOM SCHOOLHOUSE TRANSFORMED INTO A COZY HOME

Chapter Twenty-Three

John and Elizabeth Schoonmaker

Most of us have driven past these tiny charming buildings sitting so pertly in the midst of farmland. The one-room school is a relic of our bucolic past, and only a few are still in use. What becomes of them? Adaptive reuse has been the answer for some lucky ones; their imaginative owners have retrofitted them for new lives as petite country homes.

This example is in New York State, in the Cooperstown region. First erected on farmer Albert G. Huntley's land in the early 19th century, the Huntley Corner's School has had four incarnations before its present one as a country home.

In 1849 a schoolhouse costing $250 was erected on the foundation of an earlier structure. (Each local resident furnished one half cord of wood to heat the schoolhouse in winter.) The next schoolhouse was built only a few years later (1855) when the building burned; the cost had gone up to $311. (There were twenty-nine students, and they are known to have had 167 books.) Fire also claimed this building.

The present structure was erected in 1904, again on the same stone foundation and looking remarkably similar to its predecessors. This version of the Huntley Corner's School was a similarly simple, compact structure with a row of maple trees on the lawn. Boys and girls were separated by a center partition. In 1918 plumbing had been added to the building, and in 1930 a flagpole was erected. But by 1932 only eight children were attending classes in the schoolhouse.

Historic photo of the schoolhouse with teacher and twenty-one pupils.

And some years ago when the new owners, do-it-yourselfers John and Elizabeth Schoonmaker purchased the schoolhouse and surrounding acre of land (for $3500) birds were living in it. Nor were they the only problem. The foundation—after close to 150 years of use—needed to be replaced. The roof leaked. The well had to be abandoned. The schoolhouse needed both a new chimney and fireplace.

But there were wonderful compensations. The original windowpanes were salvageable. There were 10 foot high ceilings, and with the partition gone, a big airy space. The fine wainscoting remained and only required cleaning with oil soap.The floor merely required sanding.In addi tion, the owners put in insulation, built a new kitchen (where the girls' section had been) and made the loft into a bedroom.

When they outgrew the schoolhouse and sold it to its present owners, musicians Christine Ims and Michael Levin, the Schoonmakers built a new house just down the road. That way they still get to see the charming little building they saved behind its row of maple trees.

Above: Schoolhouse as it looked when purchased for conversion, with fading white paint, boarded-up windows, and leaky roof. Birds were living inside.

Below: Schoolhouse, still beautifully situated in farmland, sits behind its rows of maple trees. Here it gets a new foundation.

New chimney, new coat of red paint, new staircase to loft bedroom, and cozy fireplace adapt the schoolhouse to use as a petite country home.

Chapter Twenty-Four

A TIN-ROOFED SHEEP BARN CONVERSION IN VERMONT

Peter Kurt Woerner, Architect

Recent photograph shows the brand new "envelope" enclosing the historic frame of the Vermont barn at its hilltop site.

While many architects retrofit an aging barn by building a new frame within the old structure, architect Peter Kurt Woerner of New Haven, Connecticut, did the reverse: he enclosed the old frame within a new "envelope." The frame of the original weathered, tin-roofed abandoned sheep barn was still sturdy, but the foundation and the siding were decaying. Built in Vermont in the 1830s, the barn had an unusual construction well worth saving: an extraordinary continuous 40 foot ridge beam ran the length of the barn and the intricate, perfectly symmetrical interior frame was supported by 16 foot tall hand-hewn gunstock columns.

Woerner began by clearing away 3 acres of dense woods encroaching on the barn. The structure needed a new foundation, so instead of working beneath it, he built atop the adjacent hillside and moved the barn onto it—giving the barn a lower level and a broader view. The sloping site enabled him to add a new bedroom and garage wing on the lower level. Before rebuilding, he removed the original weathered siding. Then he constructed a new "envelope" (which indeed looks new) around the original structure, but which leaves the entire complex frame visible within the barn as a kind of free-standing structure.

In the interior, the original wood roofing boards and hand-hewn columns and beams give the 27 foot ceilinged living room a look that is both rustic and spaciously elegant. New windows stretch across the front. There is a large stone fireplace, and above the living room, an enclosed sleeping loft. Rustic furnishings include a ladder and an old wooden sled (now the coffee table), but what really furnishes the barn are the original wood rafters and beams, and a sense of old-time craftsmanship.

Above: The barn frame, with its original rusted tin roof, being moved by truck to its new, larger foundation atop a hill nearby.

Left: Interior of living room has original hand-hewn beams and columns, soaring overhead space, and rustic elegance.

A NEW TROMBE WALL MAKES
A CANAL HOUSE LIVABLE

Chapter Twenty-Five

John Ringel and Greg Torchio of Jersey Devil, with Jim Webb

Sometimes the retrofitting of one major aspect of a structure makes the difference in adaptive reuse. In the case of an old stucco-on- stone canal house (dating to the early-to-mid-1800s) in Lambertville near the Delaware-Raritan Canal system, the structure had the charm of the antique, but needed insulation and heat in order to be fully used. The building, which was known to be one of the oldest in the area, consisted of only two 16 x 16 foot rooms—one above the other (and a 1940s addition in the back dating to another life as a social club). The owner soon discovered that even very thick walls do not insulate satisfactorily.

John Ringel and Greg Torchio of Jersey Devil in Stockton, New Jersey, with the help of the owner—ceramics artist Jim Webb—undertook the project. They decided that what the canal house needed was a Trombe wall to provide solar heated air on its south side, along with insulation (sprayed in place foam) on the other three sides. Insulation was duly added on the exterior, and stuccoed to match the original finish. The additional thickness created by the insulation provided the airspace for the south side.

Canal House retains original look but has a new solar heating Trombe wall installed on the south side.

Here the major addition was the Trombe wall, which was actually a double wall acrylic system over black stucco, attached directly to the original stone side of the building by means of an aluminum glazing system. The designers left two small vents at the bottom; they can be opened in the summer for natural solar assisted ventilation. In what was once an attic space there is another outlet for solar heated air. Other major renovations included the removal of the second floor ceiling and the creation of a cathedral ceiling instead.

The addition of the Trombe wall and insulation has made the canal house usable as a live/work space for Webb, who created much of the interior furnishing, including a tile wood stove to supplement the solar heating on the coldest days.

Left: Studio interior includes a tile wood stove (designed by the artist who lives there) to supplement solar heating on the coldest days.

Below: Second floor with a cathedral ceiling has a gable vent converted to a hot air intake from the top of the Trombe wall.

MACHINE SHOP IN CHICAGO IN A SURPRISING ADAPTIVE REUSE

Chapter Twenty-Six

Kenneth Schroeder, Architect

Above right: Exterior view pictures a modest brick and white trim building on a Chicago street, with a new rooftop addition.

Above: Work in progress photo suggests the former machine shop as it was.

A particularly successful example of reuse is the conversion of a rather ordinary smallish-looking brick building facing the street in Chicago. Used as a machine shop, it actually had a sizable interior (25′ x 118′), high ceilings, a skylight and a flat roof. It became as surprisingly stunning living space under the guidance of architect Kenneth Schroeder of Schroeder Murchie Laya Associates in Chicago.

The architect decided to carve out one end of the building to create a courtyard. He also designed a small rooftop addition to the two floors of the structure, while retaining the original skylight and adding two more, surrounded by exposed ceiling beams. With the use of glass block at one end of the building he added yet more light to the interior. In the very sophisticated, spacious living room/dining room/kitchen, a large industrial window divides a wall stripped to bare brick. The kitchen area is a modern stainless steel, with black and white surfaces, while a striking black on white staircase adds to the contemporary design.

The element of surprise can be said to characterize this retrofitting. For, in the case of this former machine shop, the modest exterior gives little indication of the up-to-date elegance within.

Left: Modern design of the staircase and exposed beams add to the contemporary character of this converted space.

Below: Sophisticated interior features a huge living area with a bare brick wall, skylights, industrial windows, and glass block.

CONVERSION OF A STONE AND SHINGLED GARAGE FOR A SCULPTOR AND FAMILY

Chapter Twenty-Seven

Theodore Kessler, Architect

Before conversion: Gambrel-roofed stone and shingle garage with studio loft.

Most older houses have an unattached garage—often a prime object for conversion. Located on a hillside in a small town in New Jersey, this stone and shingled 1915 garage with a gambrel roof also had an upstairs that was used for many years as a painting studio by the owner's father. Not long ago the owner, a sculptor, decided to convert the garage into a home and showplace for his own work. He asked architect Theodore Kessler, of Leonia, New Jersey, to draw the plans; he intended to do most of the construction himself.

The garage had 400 feet of space on each level. The architect designed a two-story addition (with matching gambrel roof lines) to be set at an angle to one side of the original structure. There is a resulting triangular space in the center, which became an enclosed atrium for showing sculpture. A deck high above the ground wraps around two sides. Tall, arch-topped windows bring in lots of natural light and look out on the deck. Doing most of the carpentry himself, the owner built a double level home for himself and his family with the original structure becoming one wing of a much larger building.

Above: New addition was placed at an angle to the original structure, creating a central atrium space for showing the owner's sculpture.

Right: Interior has wood trim and new windows looking out on deck.

Below: Second-story wrap-around deck was added to two sides of the house.

THE CONVERSION OF A 19TH CENTURY LONG ISLAND CHURCH

Chapter Twenty-Eight

Guy Ladd Frost, Architect

In Roslyn, Long Island, an 1851 Greek Revival church had a typically checkered history before its adaptation. Thought to be the work of the Long Island carpenter-builder, Thomas Wood, it was constructed for a Presbyterian congregation who paid him $1900 to build it. (Among the congregation was the noted poet William Cullen Bryant—whose own delightful Cedarmere house and gardens are nearby.)

The basic design of the church was simple. It was a shingled rectangular space with two tall double-hung windows on each side and a central hall divided with two aisles. But it was augmented with a pretty steeple and a belfry, a covered porch, a choir loft supported by cast-iron columns, and fine Greek Revival moldings.

By 1928 the burgeoning Roslyn congregation outgrew its small church and—like many rural churches—it faced an uncertain future. Its center area was divided up and the building passed from one use to another. Along the way its charming steeple was lost, and it was never replaced.

Architect Guy Ladd Frost of Roslyn undertook to turn the church into a private home centering on the original 33 by 23 foot space. The ceilings were 18 feet high. Removing the partitions, he was able to recreate the great spacious area that once held the congregation. The choir loft was made into a sitting room, while the kitchen was fitted into the new vestibule. The balcony above the original doors led to a new addition of bedrooms and bath. To the side, where once horses were tethered, there are now a shingled addition and a garage, almost disguising the original church's form.

But inside, the high, open space and proportions are clear reminders of the 19th century church. The original plastered ceiling and walls, fine pine wainscoting and flooring, and carefully refinished woodwork and detailing give the house an antique quality that is reinforced by the traditional white and putty color scheme, 19th century wall decorations, and the inclusion of one of the original church pews.

Early picture of the church shows the enclosed porch, steeple with belfry, and long stable area for congregants' carriages.

Left: Today, though the original church building is quite disguised by the loss of its steeple and new additions, elegant windows and fine proportions of the original structure suggest its former life.

Below: Inside view of the spacious main room with its balcony, fine woodwork, and 19th century decorations show how antique rural church buildings can lend themselves to today's quest for large open living spaces combined with historic architectural details.

IN PROGRESS: A BARN WITH AN EXCEPTIONAL HISTORY IS RETROFITTED

Chapter Twenty-Nine

Joseph Levine, Architect

The small barn had a special place in 19th century history. The Underground Railway is known to have used the structure in eastern Pennsylvania as a hideaway. Built near the upper part of the Delaware River about 1850, the barn was constructed in post-and-beam style, with a bluestone foundation.

When its conversion to living space was undertaken by architect Joseph Levine, of Bone/Levine Architects, in New York City, his aims were to preserve the "irreplaceable structural timbers," to "acknowledge the tremendous valley view through the use of an operable screen-wall," and to honor "the structure's spirit" as inseparable from its landscape.

He was able to save the post-and-beam timber structure (which still supports the roof) and the bluestone from the foundation, while a new enclosure was constructed around the barn, "allowing the original wood structure to maintain its own identity." Red-framed windows with large exterior sliding shutters (built of barnboards) provide river views when open, and slivers of light when closed—"a gesture to the rays of light that originally penetrated the slats in the barnboard siding."

The architect decided to retain the barn's interior as one large room accommodating living room and open kitchen (only the bath was partitioned). There are hanging sleeping lofts in the bays on either side of the center space. Passive solar heating and in-floor heating, as well as a wood burning stove are planned. With the post-and-beam structure intact and its large, open interior space, the structure will clearly retain its "identity." "This project isn't the renovation of a barn, but rather the reincarnation of the barn into a house," commented the architect.

Historic barn in eastern Pennsylvania was once a stop on the Underground Railway.

Exterior today: red-framed windows and sliding panels made of barn boards (here shown partly closed) bring in light; new foundation uses original bluestone.

Work in progress: interior of barn retains original post-and-beam structure.

Interior is one large room; it will have sleeping bays above. Barn board window shutters are a gesture to original slatted barn walls showing light between the boards.

PART 2

MULTI-FAMILY AND MIXED-USE CONVERSIONS

A NEW ORLEANS SPICE FACTORY TAKES ON A NEW IDENTITY

Chapter Thirty

Kenneth Schroeder, Architect

In the heart of New Orleans, near the French Quarter in the Faubourg Marigny Historic District, stood the ruins of three buildings that had for years housed a spice factory and warehouse. The trio of low structures, which were built about 1930, formed three sides of a courtyard. Despite their rather ordinary look, the spice factory had a certain renown: famed chef Paul Prud'homme mixed his Cajun spices there.

When architect Kenneth Schroeder, of Schroeder Murchie Laya Associates in Chicago, with Brian Perry and Charles Wilson, undertook the adaptation of the buildings into 14 residential units, they had the neighborhood in mind, as well as the details of renovation. "The archeology of the former place along with the new interventions were (to be) read within the rich text of the neighborhood, site, and building details," Schroeder writes.

In keeping with the New Orleans vernacular, he conceived of the conversion as a "courtyard typology" with "balconies, walkways, and overlooks weaving all levels and parts of the complex together." In fact, the retrofitted buildings have a quaint but contemporary ambiance, in which all of the spaces are different, and there are no typical floor plans. The sequence of spaces, as he sees it, culminates in a pool adjacent to the ruin of the former boilerhouse.

He began by blocking off the brick courtyard with an entry gate and creating three interior courtyards (one of which houses the pool). He saved the exterior brick walls, but made major changes within. One building has a long white covered balcony overlooking the original court, giving a thoroughly old-world Louisiana feeling to the complex. Brick walls and open beams resting on original rough stone foundations, and aged arched doorways and passageways, also add character throughout the exterior of the complex. This project is filled with

"Before" view of the spice factory and warehouse near the French Quarter in New Orleans.

Above: Courtyard configuration of the renovated Mandeville Wharf Spice Factory Lofts features variety in sizes, spaces, colors, and materials, with one building a strong terra-cotta color, and another accented by a long white balcony.

Right: One of three interior courtyards creating quiet space amid former industrial buildings.

unexpected nooks and spaces, as well as bright colors; one exterior wall is a strong terra-cotta color. On the street-side, in keeping with its neighbors, the more dignified façade is ornamented with a wrought-iron gate entry and balcony and decorative half-round topped windows.

Interior units have a contemporary feel with a variety of angles and ceiling heights, hardwood floors and stark white walls. Large industrial windows provide natural light. Now known as Mandeville Wharf Spice Factory Lofts, this complex combines the charm of an almost Parisian enclave with the industrial loft spaces and open structural design that are in such great demand in today's American cities.

Left: Entrance gate to the street is wrought iron, in keeping with a traditional New Orleans neighborhood; renovation retained old brick walls.

Below left & right: Elegant, contemporary interiors of units are light and spacious; no two are alike.

A LANDMARK NEW YORK CITY CHURCH CONVERSION

Chapter Thirty-One

Stephen B. Jacobs, Architect

One of the finest examples of Greek Revival architecture in the city (protected by Greenwich Village's Historic District) was the Presbyterian Church built to the design of Samuel Thomson in 1846. Its imposing façade with its hexastyle temple front with six massive white Doric columns was a beloved landmark in its neighborhood. But how to save it, and for what use? The preservation of the church engendered much debate.

The solution was devised by the New York architect Stephen B. Jacobs, of thestephenbjacobsgrouppc, who came up with the plan for Portico Place, a fifteen unit condominium conversion. They were not allowed to change the structure's height or classic features. Their idea was to preserve completely the exterior, as well as to save as many fine interior details as possible—while transforming the interior into living space.

The preserved, but restored, Greek Revival façade of Portico Place, the landmark, former Presbyterian Church in an historic district of New York's Greenwich Village.

"Before" photo of the church interior with its tall stained glass windows, great height, and balcony.

The dignified façade was cleaned and repaired and left architecturally unchanged. Its original doors were restored, closed and bolted, "out of respect for the cultural milieu that the church symbolized." Entrance now is through a small, flagstone-paved mews separating the church from what was its manse (built in 1846 and also now restored). Bluestone paving at the sides of the church was also retained.

The interior, which had been damaged by fires in 1855 and again in 1902, still had lovely wood floors, wooden roof trusses, and interior cast-iron columns—features the architects wanted to save. The building was divided into two zones with concrete, fire resistant corridors separating them. This allowed the fine wood floors and massive timber trusses to be retained. The conversion divided the church's main sanctuary and attic into four floors and fifteen apartment units.

A major problem was lack of light, since the church had no windows at the front or back. The only place to bring in natural light was at the top of the building; matching the planes of the gabled roof and aligned with the windows below, new attic windows and skylight were added. Since there were windows only on the sides of the building, apartments were designed to face them, with living areas near the windows, and kitchens and baths in the interior sections. Stained glass was replaced with clear glass, to add additional light.

Each unit has a different configuration, ranging from 1200 square feet to 2100 square feet. Many have two stories and most have 15 feet high ceilings and their own balconies. And of course, what makes these apartments particularly distinctive are the elegant features preserved from the original structure: handsome, bracketed window valences, deep-toned exposed wood trusses, fine old floors, and ornamental pilasters. These details of 19th century architecture both compliment —and contrast with— the contemporary design of the apartments in Portico Place.

Right: Architect's drawing of transverse section: there are now four stories containing fifteen apartments, many of them duplexes.

Below: Interior of an apartment in the transformed church, featuring original trusses, and newly added skylight.

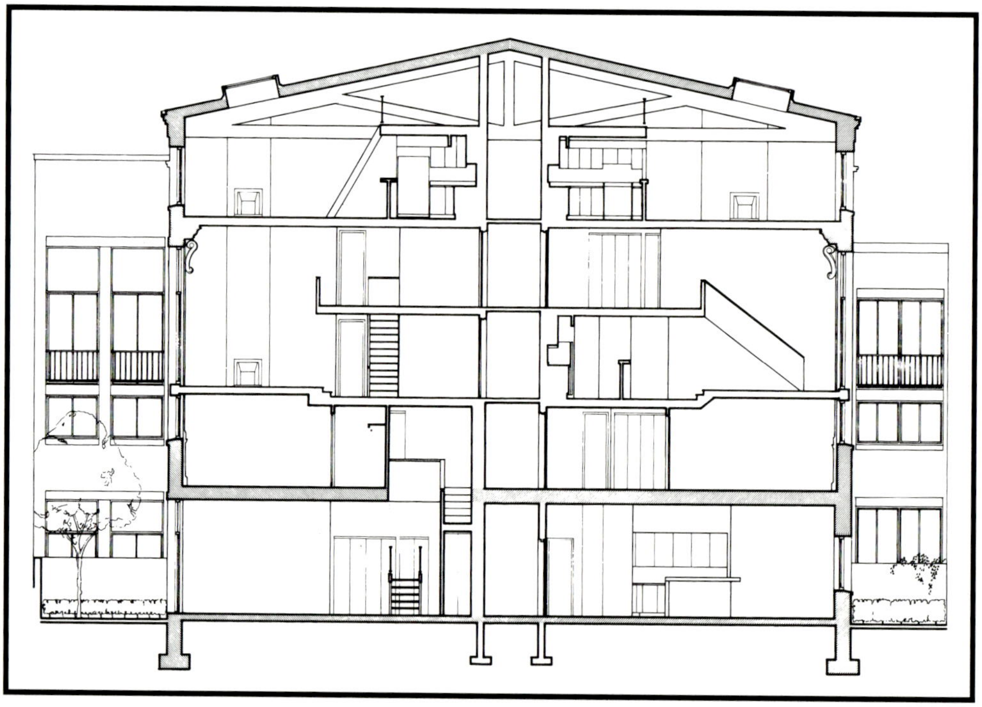

FROM SEATTLE FIRE STATION #25 TO APARTMENT HOUSE

Chapter Thirty-Two

Frank G. Stickney, Architect

A wonderful old photograph of Seattle's hillside Fire Station #25 shows three old-fashioned engines and their crews posed in front of the elegant bays of a distinctive building. The firehouse is brick with a metal tiled roof, arched windows, and harmonious proportions. It was noted for its terraced equipment bays. It was the first of Seattle's fire-houses to be built of brick rather than wood. But in 1970 the attractive brick building was in deteriorating condition, no longer viable for a modern fire company. It was in a downtown commercial zone, but as it belonged to "Historic Seattle," it was sold for development into an apartment house with many restrictive covenants. Station #25 is on the National Register of Historic Places.

The transformation of the fire station in downtown Seattle into an apartment building is a prime example of adaptive reuse. Its exterior design was well worth saving, with its four segmental-arched bays and its vast interior space. As can be seen from the circa 1920 photo, each bay opened to allow a horse-drawn fire wagon to make a quick getaway. The area in front of the station was originally paved. Behind the firehouse were stables, barn, and horse stalls. The building also had a blacksmith shop from which the Fire Department's smithy made the rounds of other firehouses in the city. In the interior there were great high ceilings and a hayloft—suggesting contemporary living spaces to today's architects.

Frank G. Stickney of Stickney and Murphy Architects in Seattle undertook the conversion of the firehouse. Retaining the distinctively arched façade, they took away the old firehouse doors and replaced them with contemporary residential doors, leaving the arched windows above. In front of each entrance they added a small fenced-in patio. Structural and earthquake improvements included floor to wall ties and diaphragms, bracing of the parapet and tower, and attic and roof reinforcement. The original metal tile roof was redone and new dormers were added to match existing ones.

In the interior the architects created 16 apartments. In the first floor space where the fire engine bays had been, the architects made use of the unusually high ceilings and attic space to create 8 units with large living rooms and sleeping mezzanines. Each apartment had between 765 and 1200 square feet of floor space on the ground floor. Second floor units had bedrooms with small decks on the garret level. The hose tower became the elevator shaft. Kitchens and baths were clustered toward the center of the building. Parking and service areas were provided in the partly excavated basement.

Firehouses like Station #25 are often among the best built and most accommodating of structures for adaptive reuse. Large city firehouses like this one, or smaller more intimate buildings that housed only one engine (like that on page **xxx**) (***Layout: this page number will need to be added as the book is put together. Thanks.**), suggest that more and more of them are being saved and reused for distinctive housing.

Above: "Before" photo, taken about 1920, shows the large bays that first accommodated horse-drawn wagons and later, motorized engines.

Right: Building section shows how space was divided.

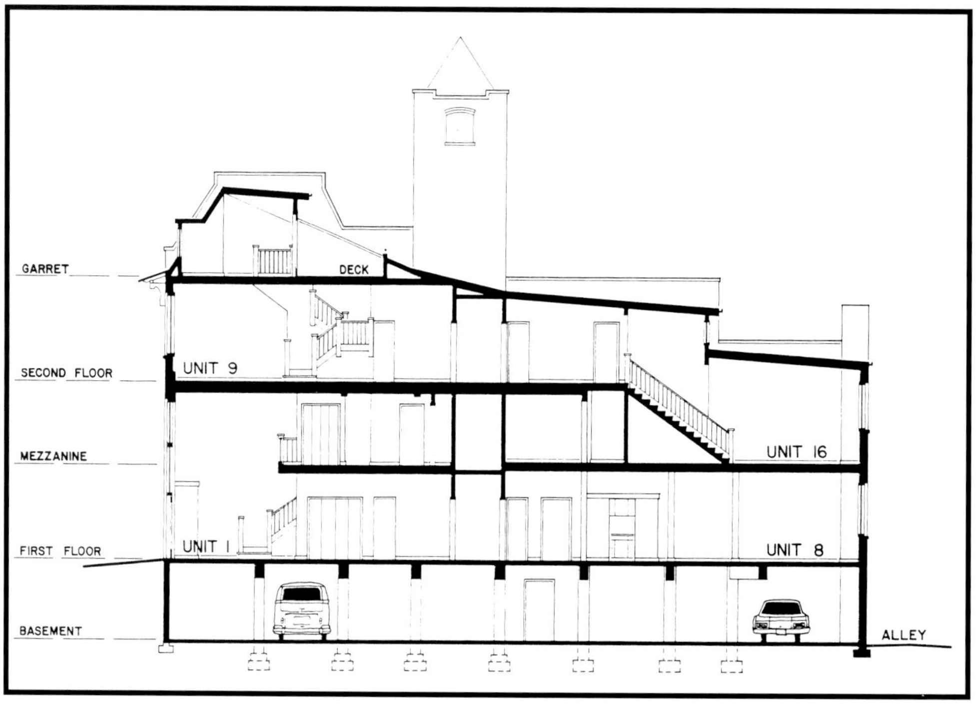

Left: Interior of one adapted apartment unit has a bricked entrance bay seen from above in the mezzanine bedroom.

Below: Converted building retains the arched bays and exterior characteristics, but a new look is created by neatly grassed-in areas, fenced-in patios, updated brick masonry, and new contemporary doors.

Chapter Thirty-Three

THE TRANSFORMATION OF A HUGE NASHVILLE SHED

Tuck Hinton, Architects

On a narrow slip of land along the Cumberland River in downtown Nashville stands a striking contemporary complex of structures housing 150 apartment units. Located in a depressed area of light industrial operations, the stunning silhouette of the new complex immediately becomes the neighborhood's eye-catching landmark.

The original shed was set on a steep riverbank, parallel to the water and overlooking a bridge. Railroad tracks and underground utilities crisscross the property. In addition to the shed structure, there was a 180′ freestanding brick chimney, a remnant of a large municipal incinerator. City records mention various use of the shed as an icehouse, granary, and gunpowder manufactory; in the early part of this century an ironworks occupied the site.

The architectural firm of Tuck Hinton Architects of Nashville turned the extraordinarily long, narrow (75′ x 900′) shed into the setting for housing units with a striking combination of preservation and dramatic new construction. The shed's "overpowering length, repetitive free-spanning trusses, and forceful axis paralleling the river" became the foundation and catalyst for the design concept.

Retained (but refurbished) is the striking steel framework of the shed's roof. The chimney was also preserved as an element of the overall design. These elements are both visual and psychological reminders of the past in this riverside neighborhood.

An entrance courtyard is created at the southern end of the property by a new "gate house" that "conforms to and reinforces the profile" of the adjacent shed. Within the shed, "an urban street" is created by the placement of a masonry screen "arcade." Under the elongated shed are three levels of apartment units, which overlook the river. The visual axis of the shed comes to an end with the second collection of apartments and a residential courtyard at the northern end, placed so as to "respond to the river, chimney, bridge, and dramatic views of the city." In fact, "framed views" of the river and city silhouette are specifically created throughout the complex by the elements of this design. To contrast with the industrial steel of the roof trusses, a warm buff masonry block was chosen for the exterior. New segments of the design are simple, contemporary, and sophisticated.

Within the new apartment complex there are many reminders of the building's industrial past. Outdated transformers have become "silent sentinels" which frame the river view at the entrance court. Salvaged metal catwalks accentuate (and contrast with) the masonry. Metal handrails and industrial fixtures have been integrated throughout; the new housing unit at the end has open steel trusses and balcony railings in the front to reiterate details of the original architecture. This converted complex, now known as the Riverfront Apartments, suggests how ingenious adaptive reuse can revitalize an aging industrial neighborhood.

Riverside Apartments overview shows how three stories of housing units fit underside of original shed, creating a striking new silhouette.

Top: "Before" photo shows exceptionally long shed in its setting along railroad tracks.

Center: Shed's iron framework offsets new housing tucked alongside it.

Bottom: Reminders of an industrial past, like the catwalk and open iron trusses, contrast with contemporary design and masonry surfaces of the new apartment units.

Above & Right: End views of the complex show a 180 foot tall brick chimney and newly built contemporary units that complete the axial design on one end. The other end has a new housing section that features open steel trusses relating it to the shed. Old transformers frame the river view.

Below: Interior of one unit overlooks wooded riverbank.

LOFTS TAKE SHAPE IN HISTORIC SAN FRANCISCO CLOCK TOWER

Chapter Thirty-Four

David Baker, Architect

These are not just any lofts in an old industrial building. They are not for the architecturally faint of heart. Each space within the building is a design statement, a melding of unusual angles, colors, shapes, and materials. Ranging in size from a gigantic 5000 square feet to a cozy 600, they have brilliantly colored concrete columns, steel stairs, undulating mahogany doors, and design details provided by local artisans. The award-winning conversion design was by architect David Baker of San Francisco.

The location, in the South of Market area overlooking the Bay Bridge, now known as SOMA, is increasingly an area of choice for live/work lofts. Within the units, bare brick walls and concrete columns contrast with the striking design components. There is a penthouse tower loft unit 15 stories above the street, with a panoramic view of the Bay and the San Francisco skyline.

One hundred and twenty-seven units were constructed within the gutted shell of the 1907 Max Schmidt Lithograph Corporation, which for most of the century turned out the classic fruit box labels used by California produce companies. The original structure was actually three separate concrete, steel, timber and brick buildings, one of them adorned with the 170 foot high, steel-framed landmark clock tower (in which the Schmidts once had a private beer hall). With the addition of new construction on one section of the roof, the total space of the project was a gigantic 230,000 square feet.

The Clock Tower building today, with its updated exterior and new construction on the rooftop.

Arranged around three landscaped courtyards that were within the original volume of the building, the lofts and the common spaces are notable for their contemporary look. A labyrinth of passageways connects one area to the next. ("Modern cities are too rational," comments the architect. "It's nice to lose your way once in a while.") Steel stairs and catwalks, angled struts, bare beams, and colorful detailing make the courtyards and common areas striking; in one courtyard, tall palms and high concrete posts supporting a catwalk are a contrast in textures and landscape design.

High ceilings, large areas of glass, asymmetrical shapes, creative room divisions, and suspended sleeping lofts are among the elements that give the units their architectural identities. (There are 44 different floor plans.) Bared brick walls of the oldest structure have been retained, and original exposed concrete columns have become elements of the design. The only completely enclosed rooms are closets and baths. Details by local artisans include the decorative metal stairway, asymmetrical lightweight concrete sink tops in all of the baths, and an unusual row of handcrafted mailboxes.

In this and other conversion projects Baker has explored "the spatial experience of a loft space." He says he has sought to design live/work lofts that are appropriately unconventional for people "who reinvent lifestyles"—much as he enjoys reinventing architectural spaces.

Right: Baisc elements of the construction were retained in conversion.

Below: Stair to the lobby area. New design features intricate division of spaces and handcrafted elements made by local artisans.

Above left: Interior courtyard has concrete posts supporting a catwalk, and tall palm trees.

Above right: Staircase area to the second floor shows an architect's delight in unusual materials, colors, and design.

Left: Typical loft interior features bare brick walls and asymmetrically shaped room division.

AN ABANDONED MEETING HALL BECOMES A "GENTLE, DISCREET" APARTMENT HOUSE

Chapter Thirty-Five

Peter Kurt Woerner, Architect

The flat-roofed, brick meeting hall was vacant and deteriorating when New Haven architect Peter Kurt Woerner undertook its conversion. Used by the Veterans of Foreign Wars as a meeting hall since 1948, the building fell into disuse while the surrounding neighborhood began attracting homeowners to its Victorian houses. Its neighbors described the shabby, single-story building on a street of Greek Revival homes as an eyesore, with its crumbling façade, boarded-up windows, trash-filled front yard, and general air of decrepitude.

When the buyer suddenly backed out, the architect—with plans already drawn—bought the building himself. Then he began work, with an eye to transforming the exterior to complement neighboring buildings and to adapting the interior into four apartments. Woerner, who described the adaptation (for which he was simultaneously owner, developer, and architect) as a "hands-on experience," was aiming for an economical route to fine design. (His solution won him an A.I.A. award.)

"Before" photo shows original boxy, flat roofed, red brick building.

He began with the exterior. Deciding that the hall would fit more naturally with its neighborhood houses of steeply pitched roofs and Greek Revival detail, he chose to add to the façade A new, low-pitched gable was added in the front to match roof lines nearby. A portico was a new addition to the entrance. Scalloped picket fencing all around (with a small pergola behind the building) helped bring the structure into the scale and era of its neighbors. Private courtyards half a level down and parking areas were also added.

In the interior, what had once been a 17 foot high meeting room is now divided into main levels and lofts. Large modern windows have been added to the rear wall for light into the living and dining spaces on the main floor and lofts above. Interiors are simple, airy, and very light. There is easy access to the outdoor courtyards.

The jury that chose Woerner's project for an award described the finished conversion as a transformation from an eyesore into "a gentle, discreet, simple neighbor."

Left: Photo of the front of the converted building shows a new pediment and portico, and a new white façade.

Below: Typical interior is a large, airy space with a loft above.

Back views of the building look more modern, with large windows; but a scalloped picket fence and decorative pergola suggest the past.

ACADEMY KNOLL: FROM CONVENT SCHOOL TO SENIOR HOUSING

Chapter Thirty-Six

Finegold Alexander & Associates, Inc., Architects

Among the aspects that make giant old buildings most appealing for adaptive reuse for multi-unit housing are aesthetically unusual architectural features that can be retained to give the new building "character." Structures built for religious use are often among America's more distinctive landmarks. Saint Ann's Academy in Marlborough, Massachusetts, an imposing brick structure, was one; it had distinctive carved dark woodwork, high ceilings, and large, interesting spaces within.

The architectural firm of Finegold Alexander + Associates, Inc., well known for their work in preservation in the Boston area, undertook to convert St. Ann's into housing for seniors. They wanted to do so by "preserving the existing character of the former Academy and its relationship to the surrounding residential neighborhood." The plan was to adapt the former school into 109 units of "imaginative elderly-oriented" housing. In addition to the structure itself, the surroundings (which include a majestic row of blue spruce in front of the building and an apple orchard, grape arbors, and gazebo overlooking a lake behind the building) have all been preserved.

The central section of Saint Ann's Academy was constructed in 1888 as a Catholic convent with thirty boarding students; in 1896 more land was purchased and an addition built. In the 1920s there was continued expansion, with two villas purchased and two new wings built. A new chapel, stained glass windows, an auditorium and additional classrooms were added to accommodate the overflow of students. The Academy continued to thrive in various religious educational roles but in 1972, after 85 years and over 12,000 students, it closed its doors.

The architects chose to retain the exterior with its decorative elements intact. But inside, they planned a dramatic overhaul, including new mechanical systems, kitchens, and baths. Individually designed apartments, with modern facilities, were integrated with historic ornamentation. Thus units in the former chapel and auditorium wing have stained glass windows and gold-leaf vaulted ceilings. The two-story chapel became four apartments. A floor was installed and some of the religious sections of the stained glass windows were removed. (Air conditioners were installed in some stained glass window openings.) Murals and distinctive, arched interior trim with delicately carved woodwork were preserved. In some former classrooms and dormitory sections, apartments have eleven-foot ceilings, wainscoted walls, and bay windows. Eleven units are designed for the handicapped. Hallways and sitting rooms feature former pews from the chapel. Residents enjoy the variation from unit to unit. "They love going to the other apartments and saying, 'You ought to see mine,'" commented the architect.

Unlike many adaptive reuse projects, this one still has strong ties to its past. With the preservation of so much of the original architectural design and religious ornamentation, residents of Academy Knoll Apartments—as it was named on conversion—are reminded throughout the building of its past.

Early photos of the Academy before expansion and reworking show the original building, a wing with stained glass windows, and original chapel before conversion to four apartment units. High ceilings were retained and stained glass windows now grace the apartments. Pews were placed in hallways and sitting areas.

Two views of apartment units: both kitchen and bedroom have distinctive arched woodwork and colorful windows.

Chapter Thirty-Seven

Rothzeid, Kaiserman Thompson & Bee, P.C., Architects

Above & below: "Before" shots of the Eagle Warehouse shows the exterior, front and back.

The original brick warehouse was built in 1894 near the old Fulton Ferry Terminal, and just south of the Brooklyn Bridge. The architect was Frank Freeman. His design featured an entry with a Romanesque arch, iron columns, a clock just below the roof, interior brick floor arches, and heavy exterior bearing walls. Within the next few years, the old Brooklyn Eagle printing plant was acquired and joined to the southern side of the warehouse (having been given new columns, six additional stories, and an added façade). Freeman also designed a reinforced concrete addition to the east. The resulting massive complex became known as the Eagle Warehouse, after the well-known newspaper (once briefly edited by Walt Whitman).

The Eagle Warehouse was used to store furniture for Brooklyn's wealthy families who exchanged their household furniture seasonally, putting away their over-stuffed winter sofas and hangings, or their lighter summer décor in the Eagle's vast spaces. When the furniture storage business failed, the warehouse served as a repository for impounded marijuana, but a successful heist ended its usefulness to the government in that capacity.

By 1979 the warehouse was no longer in use. Despite its prime, riverside location and its huge floor area, it had not been retrofitted for other uses. It had few windows facing the river and overlooking the spectacular view of Manhattan and the Brooklyn Bridge.

The architectural firm of Rothzeid Kaiserman Thompson & Bee, P.C. of

Interior "before" photos picture the condition of one storage area and fine brickwork arches.

New York City undertook to convert the Eagle into apartments. Their award-winning plan involved vital changes, though the landmark façade was kept intact. Much of the retrofitting was designed to open up the building to more light and air.

Major changes include the repeated breaking through of 20 inches of masonry in the blank waterside exterior to create hand-cut windows (including distinctive "eyebrow" windows on the top story) that would match the architectural era of the façade. Other windows were enlarged and made operable, and skylights were cut into the top floor. Even the eleven-foot-in-diameter clock in the penthouse has windows that open. A nine-story central atrium bridged by walkways on different floors was cut into the building. Some apartments are actually reached through the atrium bridges. (In the courtyard below the old "gilded cage elevator"—no longer in use—is displayed.)

Simplex, duplex, and triplex apartments, in what has now become a prime living area in the city, were created in the old warehouse. Where once furniture was stored, eighty-five different families have moved their own furniture in to stay.

Above: Renovated façade of the retrofitted Eagle Warehouse retains the original arched entrance.

Right: New central atrium is bridged by walkways; gilded cage elevator is displayed on the ground floor.

Above: Typical spacious apartment in what was once a furniture storage area.

Left: Interior of the penthouse apartment with the original eleven foot diameter clock and skylight.

FROM EARTHQUAKE DAMAGED BREWERY TO SPACIOUS CONDOMINIUMS

Chapter Thirty-Eight

Henry Siegel and Larry Strain, Architects

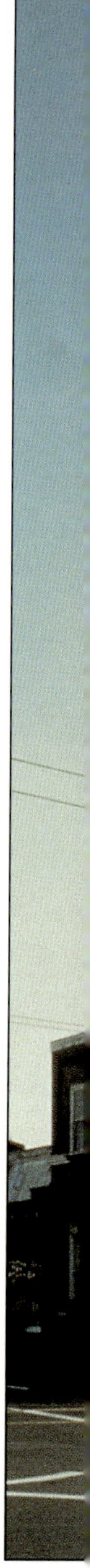

After the 1989 Loma Prieta earthquake, the turn-of-the-century Jackson brewery was dilapidated and unusable. But it had 12,500 square feet of space, and some attractive architectural features, including its hip skylight, tall round-headed arches, a tower, an arched carriage entry, and thick red brick walls above gray Colusa sandstone at the base. Its industrial Romanesque Revival style with its towered silhouette was a fixture in the SOMA neighborhood of San Francisco. Originally one of six buildings used by the brewery, it was designated a City Landmark in 1990.

Architects Henry Siegel and Larry Strain, of Siegel and Strain in Emeryville, California, came up with a mixed-use solution to the Jackson Brewery building's preservation (and won four different awards for the design). The plan called for a restaurant on the ground floor to maintain vitality in the neighborhood, and the construction of seven live/work units in the vast space above.

But the retrofitting of the Jackson Brewery was not an easy job; time (and the earthquake) had rendered the neglected building unsafe and in need of a complete reworking. The earthquake had left cracking around the sixth floor tower and along the exposed roof parapet. There was fear that damaged portions of the building would fall onto the street. And the floor of the tower had at some point been removed, leaving the tower itself unsupported for a height of more than 25 feet.

The crumbling façade was renovated with restored bricks from the building and with mortar analyzed to match the existing exterior. But the most important renovation work involved a major seismic upgrade, designed with the collaboration of structural engineer Larry Fowler. A bracing scheme was devised that involved both exterior and interior design. Its aim was to provide flexibility and to receive the impact of an earthquake, while protecting the rest of the structure. (The system contains two massive concrete columns in opposite corners of the building, with steel beams from each column running diagonally across the walls from the fifth to the basement floor, where short steel frames form the "eccentric links" of the system.) Exposed steel beams throughout the building function both as part of the system, and as design elements that punctuate the large open spaces within.

Architect's drawing pictures innovative seismic bracing scheme with diagonal supports.

Apartments range in size from 650 to 2800 square feet. The fourth and fifth floors were left in their original open configuration, while the second and third were subdivided, and the first changed to accommodate the restaurant and entry to the upper floors. The existing brick walls have been retained, structural steel frames have been left exposed, and with the high ceilings and original wood-frame windows intact, there is an overall industrial character to the apartment interiors. The combination of vast industrial space and striking contemporary details (like the freestanding steel fireplaces and open steel staircases) explains why this sophisticated—and structurally innovative retrofitting—has won so many awards.

Exterior view of the old Jackson Brewery, now seismically restructured and retrofitted for live/work lofts.

"Before" picture shows dilapidated interior with skylight and old support structure.

Above: Diagonal support system is part of the architectural design of an open, contemporary interior.

Right: Spacious interior features exposed brick walls, support beams, and a free-standing iron fireplace.

A TASTE OF TROMPE L'OEIL: AN ATLANTA WAREHOUSE WITH A SURPRISING NEW LOOK

Chapter Thirty-Nine

Anthony Ames, Architect

A derelict warehouse in an up-and-coming Atlanta neighborhood known as Little Five Points became the new home for a young couple who also placed their graphic design business in the building—and had space left over for three rental apartments. The buyers chose Atlanta architect Anthony Ames to convert the old building. He transformed the long, narrow, L-shaped structure, creating relatively straight-forward spaces for the rental units and business sections, but choosing a unique program for the buyers' apartment.

The Melia Warehouse is a white concrete and brick 11,000 square foot structure. Ames decided to divide the building into left and right sections facing the street, with the left side reserved for the owners' apartment and garage, and the right used for office and studio space. A corridor in the center of the building provides access to the rental units, which occupy the rear ell of the building.

Though his clients specified for their own apartment just "basic needs: a living room, dining space, small kitchen, bedrooms, baths, a study or gallery, and roof deck access," the architect took the opportunity of transforming the 3300 square foot area into an original and complex series of "spatial experiences" between front door and roof deck. He speaks of his plan as featuring "a variety of spatial types and experiences" including a rotunda, a vaulted gallery, and an open living area. This progression includes what the architect describes as "visual clues, such as framed views, murals, and natural light that direct (one) along this architectural promenade."

A domed skylight in a yellow rotunda provides a brilliantly light, sunny entryway. It opens into the vaulted dining room (and gallery for the owners' art). Original rough wooden posts from the warehouse contrast with a shallow sky-blue ceiling and long gabled skylight. Red oak flooring is notched to match the architect-designed dining table.

Illusion is the operative word here, for the architect has created a series of fanciful pretend windows—in fact, murals of windows—that look out on lovely views. With a flat, contemporary landscape style and latex paint, the architect himself has painted views: one is a scene of poplar trees, and another, against a rounded lozenge wall and behind a real serving counter, shows a square window overlooking a landscape of hills and sea, with a painted wine bottle and pitcher. Another almost trompe l'oeil mural decorates the curving entryway to the living room.

This contemporary-looking open space has bare brick walls and is punctuated by real industrial windows, overlooking a narrow garden. Next come the kitchen, master bedroom, and an outer stair access to the roof terrace overlooking Atlanta's skyline.

This conversion is a unique one, in that the architect has personally involved himself in both the original design and in the decorative elements—from painting the murals to designing the rug to complement its surroundings.

Low silhouette of this white brick and concrete warehouse with original industrial windows gives little clue from the outside of the unique spaces within.

Above: "Before" photo shows empty warehouse with columns (see finished dining room) and industrial windows.

Right: Architect's painting (of window with wine bottle and pitcher) before real serving counter, original support post, bare brick wall, and curving spaces are unique features of this warehouse conversion.

Spectacular sky blue ceiling with gabled skylight divides the space and provides natural light.

Left: Dining room with a painting of an illusionary window against a rounded wall, as seen from the yellow domed entryway.

Below: Curving walls, vaulted ceilings, and an open living room create a variety of "spatial types and experiences."

AN ARCHITECT TRANSFORMS A LANDMARK BOSTON MEETING HOUSE FOR HIMSELF AND OTHERS

Chapter Forty

John Sharratt, Architect

One of the oldest and most venerable projects for adaptive reuse in this collection is the Charles Street Meeting House—an architectural landmark designed by Asher Benjamin in 1807. The tall, brick, Federal style church was used by various congregations for 170 years before it finally fell into serious disrepair and became an illustrious white elephant on its elegant corner at the foot of Mount Vernon in Boston. By 1980 it had a designation in the National Register of Historic Places, as well as deed restrictions requiring that it be used only for religious, educational, or cultural purposes.

Exterior of the Charles Street Meeting House, an 1807 Federal-style Boston landmark. Renovation has left the exterior unchanged.

Debate over what to do with the landmark was finally resolved when architect John Sharratt of Boston bought the building himself with the proviso that he would preserve both its exterior and its distinctive interior architectural details if the restrictions were changed. Preservation, zoning, and a variety of other battles ensued, but Sharratt eventually prevailed. His mixed-use plan called for living space for himself and his family, his own architectural office, several ground floor shops, and two new floors with suites of offices. Among the ingenious solutions he created were exterior doors on sliding tracks that discretely vanished at night so that the church retained its historic façade, and new double-glazed windows inside the historic small-pane sash windows, preserving the historic fenestration.

The adaptation of the church was accomplished without losing the original columns, capitals, arches and ceiling decorations—all of which have been carefully restored. The front of the church from basement into the tower has become the architect's home—a commodious four-bedroom apartment; it occupies six floors of space, extending up into the tower belfry (which has a lookout through the lantern). The apartment has a private entrance on the side. The dining room and kitchen are at balcony level, while the living room is above in the building's tower. A spiral stair leads up to the library where the church's belfry once was. Below the living room in the tower are bedrooms and a weaving studio.

One of the most stunning spaces is Sharratt's double-height architectural office, which occupies four of the church's bays (but only about half the width) and has the original vaulted ceiling and great arches. Its new floor is an extension of the original balcony. The mezzanine is reached by an ornamental spiral staircase. One of the original church clocks (with its mechanism) is suspended from the vaulted ceiling, and the old church tower bell sits below it and (softly) rings the hours.

Sharratt's adaptation of the historic church was accomplished with the aim of precisely preserving the exterior, and keeping the "fabric of the interior intact." In fact, even the balcony floor could be removed. "If everything I would install were taken away," he told the Society for the Preservation of New England Antiquities, "the interior of the meeting house could become a church again."

"Before" photo of the interior prior to renovation shows the area in the church's tower that is now the living room of the architect's apartment; it has original small-paned windows and a brick arch.

"Before" photo of the church's nave and balcony; the architect's office floor is an extension of the balcony.

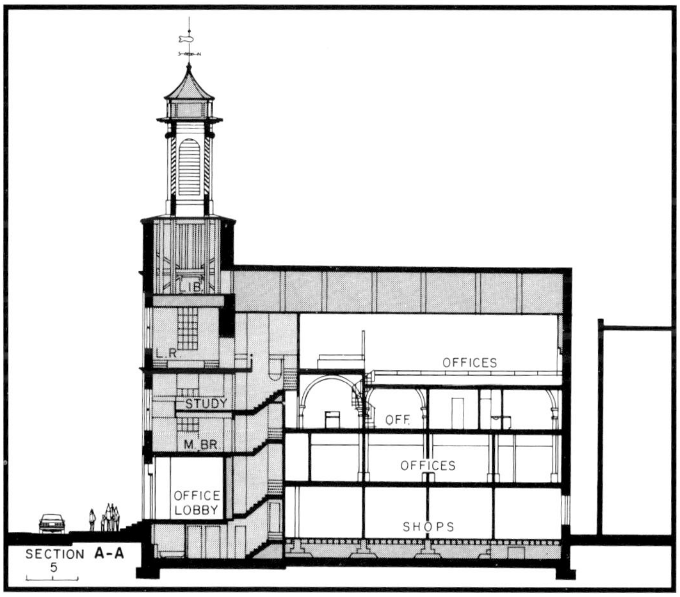

Architect's drawing of the church's adaptive reuse; living quarters and the architect's office are in the tower, while shops and offices fill the rest of the space.

Above: Stunning two-story architectural office with a spiral stair occupies part of the nave bays; original arch frames a mezzanine reached by the spiral stair.

Right: Contemporary living room in the tower features a moon-shaped doorway, spiral stair, and an original small-pained window (with new double-glazed glass inside).

HIGH ABOVE THE CITY: LIVING IN A CONVERTED CLOCK TOWER

Chapter Forty-One

Beyer Blinder Belle, Architects

Surely one of the most romantic urban settings to live in is in a converted clock tower. Not only is there a sense of tremendous height above the cityscape, but the past—even a storybook sort of past—is evoked by the idea of living behind a giant clock face. The clock tower adaptation shown here is in Brooklyn, New York. (Another, in San Francisco is on page 107.)

Clock towers were features of various large manufactories in urban areas that are increasingly becoming sites for redevelopment. As cities turn to their spacious but no longer viable factory districts for downtown living space, competing interests arise over how a city preserves its past.

The Fulton Landing Complex is in a large historic district in a spectacular location in Brooklyn, directly overlooking both the beautiful Brooklyn Bridge and the Manhattan Bridge over the East River, as well as the dramatic Manhattan skyline. Once called Gairville, for its seven industrial buildings where corrugated paper boxes were made by the Gair Company, it is now known as part of DUMBO (Down Under the Manhattan Bridge Overpass), a newly evolving area of the city. The site is a prime one for redevelopment, with some of the once gritty manufacturing district already converted to upscale living spaces. Future plans—amid some controversy—include a marina, entertainment center, hotel, park, shops, office space, and various additional living spaces transformed from former manufacturing buildings.

Distinctive clock tower skyscraper in Brooklyn overlooks the East River, the Brooklyn Bridge, and downtown Manhattan.

Among its many component parts is the 1915 Gair #7 building, a giant (275,000 square foot) skyscraper with a noted clock tower. An early example of an "all concrete" building, Gair #7 began as a vertical manufacturing building. Some twenty years ago the structurally sound building was converted to office space by the New York architectural firm of Beyer Blinder Belle. At that time, the architects turned the loading dock on the ground floor into a distinctive and elegant lobby, which was highlighted by massive concrete columns and colorful interior decorations by artist Richard Haas.

Recently the owner asked the same architects again to convert the building—this time to residential use as condominium lofts. Known for their renovations of historic sites (including Grand Central Terminal's recent transformation), the architects chose to restore the best features of the Gair #7 (keeping the lobby and clock tower) and to add new elevators, new mechanical, electrical, plumbing, and fire safety systems. They increased the floor areas of the 120 apartments by removing the enclosed exterior staircase that went up the side of the building.

All of the lofts have spectacular views, but certainly the most extraordinary is the clock tower loft itself. This "one-of-a-kind" loft apartment with four working clock windows—one facing in each direction—was recently offered for sale at four million dollars.

Above: "Before" picture of Gair Building #7 shows lower level windows and unrenovated brick façade.

Right: Four million dollar "one-of-a-kind" loft apartment in the tower features four original (working) clock windows and spectacular views.

Left: Detail of decorations to lobby originally painted by artist Richard Haas.

Below: Elegant lobby includes original massive concrete columns that once were part of the loading dock, and colorful geometric designs on walls and floor.

Chapter Forty-Two

A 1905 ST. LOUIS TEMPLE PRESERVED AND REWORKED

Richard Claybour, Architect

The B'nai El Temple is the oldest surviving temple built by St. Louis' Jewish Congregations. Designed by John Ludwig Wees (who built a number of St. Louis' well-known structures), the Temple has Romanesque proportions with a wide façade and two symmetrical octagonal towers. Three arches flank the entrance; the arches are repeated in the windows above, and in the open arcades of the towers. Elements of design—including gargoyles, terra-cotta and copper ornaments and roof finials, and detailed woodwork— show the care that was lavished on the original construction in 1905. The distinctive architecture of the Temple made it a familiar and beloved landmark in its neighborhood.

But, like so many fine buildings, it outlived its usefulness to its congregation, and by 1944 had been sold for use as a Catholic school (which converted the central space into a gymnasium), and then to the Board of Education. Finally it was boarded up, its interior left to decay. But as a National Historic Landmark, it was not torn down.

When it was sold for retrofitting as an apartment building, architect Richard Claybour of St. Louis was selected to design its adaptation into housing. Because of its landmark status, Claybour changed the exterior very little. Its red, pressed brick walls, stone foundation, and (where possible) slate roofs were restored, as were the many exterior details.

The interior, however, needed major changes, despite the efforts made to retain some of the more elegant architectural features. Among the details that survived were the distinctive banks of windows with roundels, the rounded "cove" ceilings and their arched brackets, the elegant dark carved woodwork, cathedral ceilings, and three of the original fireplaces, mantels and grates. (One fireplace was once in the rabbi's study on the first floor.)

Above & right: "Before" interior photos show the run-down condition of this boarded-up building, and previous conversion of the main temple space into a gymnasium.

Without destroying the fine architectural features of windows and walls, Claybour installed a new second floor. The building was divided into twenty-eight units; the second floor has loft apartments with exceptionally high vaulted ceilings. Some duplexes have loft bedrooms reached with spiral staircases. One apartment includes a turret. With so many giant windows, all of the units are light-filled. Efficiency kitchens are part of the open designs in the units.

Now known as the Temple Apartments, the building has begun a new life in the Shaw District of St. Louis, where it retains its historic position as a corner landmark.

Refurbished building, originally B'nai El Temple, is now reconditioned and the home of twenty-eight families. The architect retained distinctive Romanesque architectural features, both inside and out.

Right: Rounded cove ceilings with arched brackets were among the fine woodwork details preserved.

Below: Interior of a duplex unit at the Temple Apartments features original arched windows with roundels, a spiral staircase to the loft bedroom and a high, vaulted ceiling.

AN OAKLAND LEATHER FACTORY BECOMES LIVE/WORK LOFTS

Chapter Forty-Three

Philip Banta and Associates, Architects

The 1916 factory building was once used to make leather pockets, but when architects Philip Banta and Associates of Emeryville, California, came upon it, it was in severe disrepair, and its second story wall was being used by local youths as a graffiti gallery. Furthermore, it was not seismically upgraded—a requirement for buildings in Oakland. Nor was it considered to be a building either "historically or aesthetically significant." It was located in "an industrial area in transition." The worn brick structure had windows only on one side.

But it was near Jack London Square food market, an area of Oakland that is now being transformed into a true mixed-use community (including many artists). The building had lots of space (18,731 square feet of it), the windows it had faced north, and it offered an alternative to more expensive lofts across the Bay in San Francisco. The decision was made to keep costs low and the construction simple in the retrofitting, so as to offer affordable units that could be used both for living and working.

The architects decided to make dramatic alterations in the exterior. Major renovations to what is now called the "Pocket Building" included the replacement of three of the exterior walls with a new arrangement of "a deep stucco plane to hold the street." On the east side the wall was opened up completely for additional space and a galvanized metal clad "pocket" was cantilevered over a striking bent steel beam. Only one wall was saved: the party wall with a neighboring building.

In the interior the timber posts and beams were sand blasted, the existing brick walls were water blasted, and new concrete surfaced, wood framed, and detailed floors were installed. Outsized mahogany framed custom windows (which are less expensive than metal) brought in abundant natural light, offering an "ideal setting for making and showing sculpture and artwork." The building was divided into 10 units, each of about 1600 square feet, with gas kitchen, bath, sleeping platforms, and large, high ceilinged live/work studio space.

Other amenities included aniline dyed studio doors, embedded wooden handrails, and abstracted aluminum grills for the mahogany street doors. These details were designed "to give spirit to a low cost project." In one of the most unusual aspects of its preservation, some of the original graffiti was preserved and now provides an unusual visual contrast with the contemporary white, smooth-finish drywall. The Pocket Building thus provides an environment for artists that comes with some colorful walls that are a reminder of the building's past.

"Before" picture shows an old brick factory on an Oakland city street where leather pockets were made.

New exterior includes a cantilevered "pocket" atop bent steel beam. Large mahogany-framed windows provide abundant natural light to studios within.

Typical loft interior has cement floors, mahogany-framed windows, bare beam ceilings, and about 1600 square feet of space for living and working.

Above left & right: Contemporary interiors have a variety of differently shaped spatial divisions, open beams, and skylights.

Left: A most unusual aspect—and reminder of the building's past—is the retention of graffiti painted by local youths on an interior wall.

CONTEMPORARY HOUSING IN A FORMER NEW ORLEANS GYMNASIUM

Chapter Forty-Four

Kenneth Schroeder, Architect, with
Brian Perry and Charles Wilson

Architect Kenneth Schroeder of Schroeder Murchie Laya Associates of Chicago has undertaken many conversions to housing from manufacturing sites or warehouses, but surely among the most unusual was this New Orleans gymnasium. Built some fifty years ago, the gym complex was owned by a Christian Youth organization, and noted for the famous boxers who used it in the 1950s.

When the building was sold, it was in ruins. But its location just four blocks from Jackson Square in the French Quarter's historic district meant that the derelict building was in a prime location for development as housing.

Schroeder, with associates Brian Perry and Charles Wilson, have come up with an approach they describe as one of "selective subtractions and additions...coded, structured and articulated as part of past and present change." They have taken into account the "archaeology" of the former structure, as well as the "rich text of the neighborhood."

"Before" composite picture shows wreck of original gymnasium with the open truss work that became a signature of the adapted structure.

With an eye to historic New Orleans design and style, the architects have centered the adapted plan around courtyards, covered walkways, balconies, and overlooks. One courtyard is created by carving a pedestrian and auto area from the original building configuration. The original layout had a central gymnasium with an L-shaped series of conference rooms around it. The architects have retained original design elements, such as the iron truss work, the circular windows to the courtyard, and even the scoreboard—as mementos of the building's past use.

The plan, known as Stella Maris Lofts, called for twenty duplex and triplex apartments, with each one of a different shape and size, depending on the original spaces. According to the architects, "each unit evolved from the intersection and interaction of past and present." There are no typical floor plans. The project is described as "modest" in cost.

The reuse of the building has taken interesting shape. Two uncovered courtyards have been created. Open trusswork adds a contemporary look. A variety of walkways and different levels adds to the element of surprise, as does the use of sudden bright color. The gym itself has become a swimming pool (with original scoreboard) under a white latticework ceiling. Seven living units have been created in the space under the original gymnasium, while others have been adapted from the conference room area.

Floor plans, walls, windows—the very design—of each apartment varies considerably, for the Stella Maris Lofts project combines updated, but traditional New Orleans elements with what is unquestionably a taste for the contemporary.

Exterior shots show original brick facades, new doors and windows, but otherwise give little hint of the transformation within.

Typical living room has an unusual layout, odd divisions of space, and a bare brick wall.

Above & right: Variety of line, exterior walkways, interior balconies, and open tresswork characterize the Stella Maris building's reworking.

Below: A swimming pool, where once the gymnasium was, retains the original scoreboard high on the wall. Bright colors throughout the complex, like the brilliant red wall here, add interest to the interior spaces.

A WEST TEXAS FORT TAKES ON A NEW IDENTITY

Chapter Forty-Five

Chris Carson, Architect

In the rugged country of West Texas just 25 miles north of the Mexican border, amid 25,000 acres of canyons, rocky peaks, and mesas, stands an eye-catching and historic adobe building. It seems to guard the landscape, and in fact, is an antique fort once known as El Fortin del Cibolo, and now called Cibolo Creek Ranch. It was privately built in 1857 to fend off Apaches and Comanches, as well as bandits hiding in the hills in the Wild West days of the 1850s.

A West Texas cattle baron and Indian trader named Milton Faver (locally called Don Meliton) erected three such forts to protect his vast Longhorn cattle and sheep ranching and farming empire—as well as his family, workers, and private army. The major fort has a low-profiled Southwestern architectural simplicity, with adobe walls three to four feet thick, diagonally opposite round watch towers at each end of the one-story structure, a center courtyard, and a distinctive "defensive solidity." It is one of Texas' most significant historic structures.

After Faver died in 1889 (his glistening white adobe tomb can be seen on a hill above Cibolo Creek), the forts changed hands and fortunes, but continued to be used by the ranching community. A few years ago the crumbling buildings were purchased by a businessman who decided to convert the antique forts into living accommodations. The large size and spectacular design of Cibolo Creek suggested taking in guests. With that aim Cibolo Creek Ranch was retrofitted—but its ambiance was carefully preserved, using historic documentation and archaeological evidence. This rare example of a private fort is a National Historic Site.

Cibolo Creek Ranch is set in the rugged West Texas countryside. Its conversion retained the original Southwestern fortress architecture.

Above & right: Cottonwood beams are used both on a spacious, long veranda and in interior ceilings, as in the original 1857 fort. Louvered doors open onto the veranda, which overlooks an interior courtyard. Slatted roof "latillos" protect the porch from the hot Texas sun.

Architect Chris Carson of Ford Powell and Carson in San Antonio undertook the transformation, which lasted for four years. Major repairs included restoring the crumbling adobe walls of the forts with new hand-made adobe, rebuilding dry-stacked stone walls, and repairing the stone-line irrigation "acequias" originally built by Faver to bring water from the natural springs to his acres of peach orchards and vegetable fields. An additional building in a similar style was built nearby.

In his award-winning conversion Carson created eleven guest rooms in the long hacienda (with veranda) in the largest fort, and four in a smaller fort (also on the vast property but about 25 minutes away), where some accommodations are actually in the lookout tower. Part of the main fort is now a museum.

Cottonwood beams and split soffit strips (known as latillos) shelter the porches from the hot Texas sun. Although the exterior façade has only a few small antique windows, the center of the structure, which opens onto a courtyard, has louvered doors looking out on long verandas. In the interior, original ceilings, supported by cottonwood beams, were reconstructed with the aid of historic photos. There is a true sense of place in this conversion, from the overall Southwestern style of the architecture to individual details like the adobe, mesquite-burning fireplaces.

Above: Round lookout towers are built of hand made adobe, and have wooden drain spouts called "canales;" restored "acequias" from natural springs provided irrigation to vast farmlands.

Right & below: Interior rooms have a Southwestern look with adobe walls, mesquite-burning fireplaces, and simple décor.

CONVERSION OF A BARBER AND BEAUTY SUPPLY COMPANY IN DOWNTOWN DENVER

Chapter Forty-Six

Oz Architecture and Hobbs Design

Top & right: Buerger Brothers Building consists of what were once two separate structures used to manufacture and display beauty supply products. Façade of the larger building is decorated with polychrome terra-cotta designs and tall ornamental concrete nature patterns in the Art Deco style.

An increasing taste for loft-style living has begun to change the uses and fortunes of downtown Denver. Typically, old industrial and warehouse buildings are being converted to mixed residential and commercial use at—so far—affordable prices.

An interesting example is the combined Buerger Brothers Building and Denver Fire Clay Building, which were two adjacent structures on a downtown street. The Fire Clay Building, a three-story wood frame structure which dates to 1892, was used as a showroom for scientific instruments. In 1929 the larger, cast-in-place concrete frame Buerger Brothers Building was put up next door to house a beauty supply company. It was designed by a noted Denver architect, Montana S. Fallis, who was well known for his Art Deco style. The Buerger Building was a fine example of its period, with attractive polychrome terra-cotta detailing on the façade and interesting spaces with exposed concrete and hollow core clay tile masonry within. In 1939 the Buerger Brothers bought the smaller building next door to expand their barber supply division.

OZ Architecture and Hobbs Design of Denver were chosen to adapt the site and to turn what had been a large derelict space (but in a structurally sound building) into 31 loft apartments, with the first floor and mezzanine occupied once again by a commercial tenant (who has retained the original spacious layout with surrounding balcony). The upper floors—including the third floor "auditorium" (where demonstrations of hairdressing took place)—are adapted to residential lofts. On the fourth floor, where face powders were once milled, cosmetics packaged, and industrial blades sharpened, there are eight additional living spaces. Top floor units have a roof deck atop the smaller, original Fire Clay Building.

The renovation has changed the exterior sparingly, as befits an historical building with such a distinctive façade. The exterior, featuring detailed terracotta designs and tall, decorative concrete nature patterns, was cleaned and repaired where necessary, and a steel canopy with wrought iron decorations was refurbished.

In the interior, the architects kept the original columns and windows, but removed and replaced damaged walls and ceilings, partitions, and mechanical systems. Artistic details, like decorative stenciling on the third floor walls, tile masonry, and ornamental mezzanine railings, were preserved. Kitchen and bath units and bright new flooring complement the original big windows, high, exposed beam ceilings, and original brick walls. And coincidentally, the new tenant for the first floor and mezzanine is once again a beauty supply company. The conversion of the Buerger Brothers Company thus very neatly combines the old and the new.

"Before" pictures show how the interior looked before renovation: showroom floor and mezzanine featured heavy concrete columns and decorative iron railings (both of these elements were retained), while upper floors needed a total overhaul.

Two adapted spaces in the converted building have original windows, open ceiling beams, bare brick walls, and new flooring, as well as lots of open, airy loft space.

THE CONVERSION OF A REFRIGERATED WAREHOUSE IN NEW YORK'S GREENWICH VILLAGE

Chapter Forty-Seven

Rothzeid Kaiserman Thompson & Bee, Architects

A seemingly hopeless target for conversion, the refrigerated warehouse complex was a "white elephant," hardly an obvious choice to become luxury rental apartments. It was a block square of ten buildings, constructed over a period of years from the mid-1800s to 1930 in the burgeoning meat market district. The buildings had four or five different structural systems, with differently aligned floors. There were few windows or interior courts for light. Of particular note was the railroad track that came in on a trestle above the street—right through the second floor!

When the architectural firm of Rothzeid Kaiserman Thompson & Bee of New York City took on the project, the walls and windows were thick with a foot or more of frost. The interiors of the buildings, which had been insulated in Portuguese cork, were encased in thick ice. One building held the machinery that made the chilled brine

Above & left: : "Before" photos show the refrigerated warehouse with a train ramp entering the second floor, and the interior condition of the first floor loading dock.

that was piped throughout the neighborhood to keep other refrigerated buildings in the meat district cold. The entire complex needed defrosting. (The process took six weeks to prevent the walls from buckling.)

But there was a full block of space. And the buildings were soundly constructed. (The various structural systems included brick bearing walls, reinforced concrete of several types, and fireproof steel.) The architects decided that cutting a central courtyard into the building was unfeasible; instead, they cut up the area into a variety of lofts and apartments, all accessible from a central lobby and elevator system. They created long hallways of varied widths, heights, and decoration to help counteract the unusual length and depth of the building. Atop the tallest section, a modern "windmill" generates enough power to light the long corridors.

The extensive depth of the structure meant that many of the 250 units are only 12 feet wide, so they are designed as hybrids of lofts and conventional apartments, with partitions of differing heights and design. Spaces near the newly added windows are kept large and high ceilinged, to permit passage of light to the interior sections. The center portions are made into alcoves with low walls facing the windows, and higher walls facing the rear. Kitchen and dining areas are semi-enclosed at the opposite end from the windows. Where the cooling tower once was, triplex apartments now exist.

On the ground level of the newly named West Coast Apartments, a series of period arches and lighting, and newly planted trees revitalize the streetscape. As for the train ramp, it still heads from the street into the building, but it no longer carries refrigerated beef.

Retrofitted building has new windows and repeated period arches and lights along the street.

Above & left: Interior shots of one apartment feature a loft-like open space and bare brick wall; complex alcove and ceiling heights break up the unusual depth of the unit.

A FIRE-RAVAGED LANDMARK CHURCH CONVERSION IN BOSTON

Chapter Forty-Eight

Graham Gund, Architect

A prominent landmark in Boston's Back Bay section, this 1891 Romanesque Church on Beacon Street was nearly destroyed in 1978. A spectacular fire engulfed the church, leaving only two walls and the tower intact. Since the church was an historic neighborhood landmark, Boston architect Graham Gund, who is well known for adaptive reuse and preservation, undertook to save as much as possible of the original structure as he converted the property to housing. It is now known as Church Court.

The architect describes the Back Bay neighborhood as having elegant late-nineteenth century town houses that display much of the variety of Victorian domestic architecture. Into this neighborhood he added three townhouses within the church's remaining walls, and a new seven story brick building containing forty residential condominiums at the opposite corner of the site. "The massing and patterned masonry façades of the new building," he comments, "were inspired by this neighborhood, with its projecting bays, bowed fronts, and brickwork patterned in different colors."

Above & right: Fire ravaged this Romanesque Church in Boston's Back Bay section, almost destroying a neighborhood landmark.

The distinctive façade of the church, with its tower and Romanesque design, were renovated. And through the glass front of the old structure an enclosed, landscaped garden can be seen in a space once occupied by the nave of the church. This courtyard garden is formed by the remaining section of the church and the L-shaped new construction. Beneath the garden is an underground parking lot.

The townhouses within the walls of the old church retain recognizable architectural characteristics of its former use. Among the most striking are the exceptionally high ceilings with tall, arched windows ornamented above with additional small inset Romanesque windows. And most notable are the exposed wooden ribs of the vaulted ceiling, and the warm-toned wood staircase and balcony reminiscent of the pulpit in what was once the church—but now giving architectural character to a serene living room. And one apartment unit—in the tower—has seven stories.

Church Court is an example of adaptive reuse in which the preserved structure is only a part of the finished project. While the historic façade, with its magnificent rose window, three-arched entrance, and Romanesque tower, still graces Beacon Street, these features are now integrated into a complex of contemporary—but not unrelated—structures.

Restored façade and church tower now form part of four townhouses, with a new seven story building and garden behind it.

High ceilings and tall, arched windows now grace the living room of one of the townhouses. The vaulted ceiling with exposed ribs and warm-toned wood staircase and balcony are also reminiscent of the building's past.

TRANSFORMATION OF A NEWARK TOY FACTORY INTO LIVE/WORK LOFTS

Chapter Forty-Nine

Troy West, Architect

Architect Troy West of Newark, New Jersey, found the circa 1880 Dietze building on a quiet city street. It had been a factory, making small mechanical toys in its 5000 square feet of space, and with its thick brick walls and open area, it seemed like an ideal building for conversion into live/work lofts in the growing artistic community. West himself is both an architect and sculptor who planned to save one loft for his own use.

West divided the building down the middle and created ten lofts, each of them about 2200 square feet, with an average floor area of 25′ x 100′—an ample space for both living and working. At one end of the building he created a stairway and entrances to each unit. The architect was able to save the long rows of curved-top industrial windows to bring in large amounts of light to each loft. Supporting beams were left in place, sometimes stripped, sometimes painted. Some floors and ceilings were refinished; others have the original industrial beams and piping exposed. Each loft was retrofitted differently, according to the needs of the individual tenants.

In West's light-filled, high-ceilinged, L-shaped unit, he built a sleeping loft (with bookcases underneath) along the long length of the space, and created a kitchen (which includes the original old industrial sink) at one end of the room. His architectural office is in the ell at the opposite end. West's own large welded sculpture divides and punctuates the two sections.

Other lofts in the building, and his son's pottery studio and gallery where the boiler room once was, have somewhat different solutions to the division of space. In the studio of one artist, storage for paintings is built above the floor level, while in the loft of a technical designer, the space has been divided with semi-partitions.

Open and affordable live/work space, like that in factories converted to loft floors, is notably hard to come by in downtown city areas. In Newark its is a fairly new phenomenon; the Dietze Building is one of several such projects attracting artists, architects, and designers. What is particularly appealing about this retrofitting is that these lofts are unpretentious and inviting spaces; as an artist himself, West knows what working artists want and need.

"Before" photo shows the exterior of an old toy manufacturing building in downtown Newark. Exterior brick and windows were preserved.

Photos of the architect's own space before conversion to a live/work loft shows the condition of the factory interior. Once emptied, the assets of this space are evident: a vast area, repeated arch-topped windows, and exposed ceiling beams. A sleeping area above and storage along the right wall are under construction.

Architect's live/work loft has an office section in the ell with his own welded sculpture dividing it from windowed living area.

Lofts on other floors are occupied by artists, architects, and designers; their interior arrangements show how differently open space can be used.

A STONE LIVERY/WAREHOUSE IN HISTORIC WISCONSIN TOWN TAKES IN GUESTS

Chapter Fifty

David Burrows of Glueck Architects, and Jeff and Deborah Donaghue

Described in 1856 (two years after it was erected) as "the most prominent building in the Village" of Mineral Point, Wisconsin, "Brewery Creek" is an uncompromisingly rectangular three story structure with heavy limestone walls and a stark, old-fashioned exterior. The building had been constructed as a warehouse for an anticipated railway to serve the lead mines (and then the zinc works) which once dominated the area. Just a few hundred yards away on a creek behind it was Wisconsin's first commercial brewery, begun in 1835, making Mineral Point "the Place Where Wisconsin Brewing Began."

When mining—and the railroad—moved on, the building then served variously as a cold storage facility, an insulation factory, a livery, and for a long time as veterinary offices. By the 1970s it was dilapidated and was still standing only by the grace of preservationists who repaired the exterior walls and roof and restored it "close to the condition of a new 1854 building."

It was purchased not long ago by Jeff and Deborah Donaghue, who had great mixed-use plans for it; they chose architect David Burrows of Glueck Architects in Madison to design its partly commercial conversion. The new owners planned to create living space on the upper two floors, with a pub and restaurant downstairs, and—continuing the town's historic past—a new brewery. Brewery Creek, as they named it, was to become a small (four guest-room) inn, housing their family, as well as visitors to historic Mineral Point. Their apartment can also be converted to two additional guest-rooms.

"Before" photo shows desolate 1854 warehouse with boarded up windows and loading doors on the second and third floors.

The renovation of the structure was accomplished without the loss of many of its historic architectural features. Though the exterior looks much the same (though freshly restored) with its limestone façade punctuated by unadorned windows and loading doors, the interior needed massive renewal. There was no plumbing or electricity, there were no interior walls, windows or doors, and the first floor was dirt.

The renovation plans preserved the exposed heavy stone walls and timber posts and beams. In order to adapt the building to its current use, floors were leveled; the dirt floor on the ground level was replaced by concrete with in-floor heating. Stairs were added and utilities were installed. Rooms were created and reproduction windows and doors were added. So as to maintain the exposed timber and stone wherever possible, insulation and drywall were not used in the downstairs public rooms or the inn guest-rooms.

In the converted ground floor, a small, modern brewery now functions—it can make 3000 barrels of beer per year. (The restaurant offers many dishes that use beer as a vital ingredient.) The conversion of this old stone building is an example of adaptive reuse that is oriented toward the history and commerce of a neighborhood, as well as toward preserving a distinctive village landmark.

Above & left: Interior "before" shots picture original post and beam construction (over the years they rotted away), the dirt floor and original ground level limestone walls.

Updated exterior of Brewery Creek, now including living space, a restaurant, pub, and brewery.

Interior spaces capture an old-time atmosphere: bedroom has a fireplace and stone walls, while the restaurant, with salvaged antique wainscoting and bar, retains limestone walls and exposed post and beam construction.

SIX ADDITIONAL MULTI-FAMILY AND MIXED-USE ADAPTIVE REUSE PROJECTS

The projects that follow are unusual examples of adaptive reuse, or are still in the process of conversion. Each suggests the imagination of the owner, architect, or preservationist who conceived of new plans for old buildings.

Chapter Fifty-One

SAVING THE EXTERIOR OF A LANDMARK BOSTON CHURCH

Finegold Alexander + Associates Inc., Architects

Clarendon Square is now a seven-story building with a mix of studio and one and two bedroom apartments. These housing units are constructed entirely within the shell of what was once the Clarendon Square Baptist Church, an 1869 red brick Victorian Gothic building in Boston's South End. A victim of a devastating fire in 1982, all that was left was the towered façade.

The architectural firm of Finegold Alexander + Associates Inc. of Boston decided to construct a new steel frame building entirely within the shell of the original church. The new structure has a restored church façade (with new brick walls and gables, as well as vertical windows and lintels that are compatible with the original style). In some places the restored exterior has been expanded, as in the case of new copper dormers. The original 95 foot steeple has become a three story clock tower (which now contains a triplex apartment). On the opposite side a new copper cupola provides a transition from the Victorian façade to the new contemporary roof line.

And where the rose window overlooked the entrance, there is now a new stained glass window designed by a local artist. Thus, though only the exterior remains, the identity of the street façade has been preserved.

This is all that was left of the burned out Clarendon Square Baptist Church, with its Victorian Gothic arched façade and tower.

Top: Restored exterior: steel beamed new apartment building behind renovated façade.

Below: Exceptionally light apartment interior in the restored tower overlooks the city of Boston.

IN PROGRESS: FROM GREENHOUSE FACTORY TO AFFORDABLE HOUSING AND LIBRARY

Chapter Fifty-Two

Stephen Tilly, Architect

The Burnham Building has been a notable fixture in the small Hudson River village of Irvington since the 1880s. Its outlook on a town corner facing the Hudson and the railroad tracks—as seen in an early drawing—has changed little. The massive red brick towered building dominates its surroundings. (The building is actually a series of three structures—the factory, a stable or carriage house built about 1881, and a third building dating to 1950.

In keeping with its imposing size and prominence in the village, the Burnham Building has an illustrious history. It was the largest factory to produce iron greenhouses or conservatories—like that made famous by Sir Joseph Paxton's Crystal Palace in London—in the late 19th century. As iron-framed glass conservatories became increasingly popular across America, the Burnham turned out ornate, technically advanced structures—and even featured an example of a greenhouse on the building's top floor. It was apparently used for demonstrations of the new technology at the hands of Messrs. Lord and Burnham, the directors. Among their clients (and financiers, as well) was railroad tycoon Jay Gould, who had a conservatory built there for his elegant gardens at nearby Lyndhurst.

In this newspaper clipping from the 19th century, Burnham factory can be seen with train and river traffic across the street.

Architect Stephen Tilly of Dobbs Ferry has undertaken the conversion of the Burnham Building into 22 affordable housing units, with a new public library on the ground floor. The building will provide studios, one-bedroom and two-bedroom apartments on the upper three floors, for people of moderate income. The library will have 10,000 square feet of space.

The existing greenhouse on the top floor—which has 14 foot high ceilings at the center—is becoming an atrium/community room for the use of the tenants. Its roof, by the way, has a demonstration photo voltaic system; it runs the electric meter backwards, offsetting the cost of running electric shades on the greenhouse, and automatically regulating the amount of sun to shine through!

The typically American Victorian structure of the building—with its red brick façade and corner timbered tower—is being restored. But adapting the interior of the building has required major renovation. Windows in the old part will be restored when possible, while new ones matching the historic style will be used elsewhere. Flooring has been reinforced, the pitched, tin-shingled roof replaced, and the interior walls and working systems are being redone.

The successful combining of affordable housing (and financing) with a needed town facility, has made the adaptive reuse of a landmark building a reality. And the transformed Burnham Building has recently been nominated for historic landmark status.

Recent exterior photo shows little change in the red brick Victorian façade with its corner tower.

Original rooftop greenhouse, once used to demonstrate the latest conservatory technology, will become the top of an atrium for tenants of the building.

Timbered tower and rooftop greenhouse have been restored.

PUBLIC SCHOOL PROJECT I: FROM ABANDONED SCHOOL TO CONDOMINIUMS

Chapter Fifty-Three

Graham Gund, Architect

School buildings—particularly those built in the first half of this century—are among the best constructed of all public buildings, often featuring large lobbies, high ceilings, and oversized windows. Many of them have been turned into living space, with varying degrees of imagination. The two examples shown here—both in the same Boston region—are creative examples of this trend toward adaptive reuse, and each has made the common areas the focal points for the overall ambiance of the structure.

The School House Residences in Boston were converted by architect Graham Gund of Boston. The old Charles C. Perkins School was turned into twenty-one condominium residences. It was the first private enterprise of its kind in Boston. The attractive interior renovation produced comfortable, efficient units at a "reasonable cost." Each of the ten classrooms, the ground floor, and the attic became one, two or three-bedroom apartments—with varying layouts and architectural detail, including some with sloping eaves and balconies.

Central to the design was the lobby, which was transformed into "a metaphorical garden complete with stylized trees, vines, and flowers." The architect designed whimsical railings, partition shapes, and garden colors to help create a common space that, unlike conventional lobbies, "makes coming home a decidedly more interesting and inspiring experience."

The Charles C. Perkins School was converted by architect Graham Gund into twenty-one condominiums.

Above: Interior spaces have a variety of architectural layouts and details; this one has sliding glass doors and a balcony.

Right: Whimsical lobby design resembles a flower garden, with metaphorical shapes and garden colors.

PUBLIC SCHOOL PROJECT II: FROM JUNIOR HIGH TO AFFORDABLE HOUSING

Chapter Fifty-Four

Finegold Alexander & Associates, Inc., Architects

The Muzzey School, dating from 1900, is in the center of Lexington, Massachusetts, not far from Boston. The outdated school was seen as a good candidate for affordable housing, for it needed little structural work and was both spacious and elegantly detailed. The architectural firm of Finegold Alexander + Associates of Boston undertook the conversion.

A variety of units –71 of them—have been developed in the former classrooms, as well as in the old auditorium, gymnasium, and attic. A senior center is also accommodated. Units include garden apartments, townhouses, and traditional apartments with one to three bedrooms. Attic units have private roof terraces.

The most dramatic changes are the refurbishing of a 65 foot high, skylight illuminated, atrium in the former auditorium, and the conversion of the elegant staircase and rounded mezzanine into a reception area that is graced with the original wood detailing and lots of glass. Other features preserved from the past include the original arched windows, pilasters, wainscoting, cornice trim, and light sconces; these details have been artfully incorporated with contemporary elements like open steel beams and spiral staircases.

Former Muzzey Junior High School, a large brick building now converted to affordable housing.

Right: Once a classroom, this two level apartment unit has hardwood floors and original large school windows.

Below: Interior of Muzzey School Housing shows the original wood stair and mezzanine with new steel beams and a new function as a reception area.

IN PROGRESS: LOWER EAST SIDE THEATER TAKES ON A NEW ROLE

Chapter Fifty-Five

Rogers Marvel, Architects

Snapshots of Beethoven Hall's past lives: the original building, the fire destruction, the sound stage.

In 1860 it was erected by a choral society, and named "Beethoven Hall" in the bustling community of New York City's Lower East Side. Its elegant exterior had a façade of cut sandstone, decorative dentils, and large Palladian windows. A double-height lobby, cast-iron balconies, and a sweeping marble staircase created an elegant interior. By the turn of the century a next door tenement was annexed, the party wall was carved up, and larger concert and ballrooms created within the two buildings.

But with community support for concerts fading, Beethoven Hall became a film studio in 1955, with many changes to its architecture—including the cutting of giant holes in floors and walls to create access for equipment, the soundproofing of walls, and a cover of cement block for the marble stair. A catastrophic fire destroyed the roof and some of the sound stages in 1992.

The building's latest incarnation was as an avant-garde theater and gallery space: The New York Kunsthalle. After renovation, three classically proportioned spaces—without any columns—were opened. Designed to show contemporary art and performances, the Kunsthalle brought in a wide variety of installations and international shows to its high ceilinged, contemporary spaces.

In 1998, once more christened Beethoven Hall, the building is again in the process of transformation—this time to residential units. Rogers Marvel Architects of New York have undertaken its conversion to living space. Plans include a thorough modernization of the space and its division into four large apartment units. There will be two lofts (one of 6600 square feet), a penthouse with a roof garden, and in the old tenement building, a split-level apartment. One of the most original features is the bow-string, glue laminated wood truss roof which will grace one apartment. The building, which will have a total square footage of 36,000 square feet, will also include a commercial space at street level.

The former hall is being divided into four large apartment units. The new arched bow-string roof in this apartment is one of the completed elements.

THE MOST UNUSUAL CONVERSION: A TROY GAS TANK GETS A NEW USE

Chapter Fifty-Six

Joe Lazare

Great blue converted gas tank overlooks a Hudson River marina. Outside stairs and balconies provide access to its two upper stories.

Surely one of the most unusual conversions of all is this huge, round former gas tank. Adjacent to a marina, it sits along the Hudson riverbank not far from Albany, New York. Converted to multipurpose use, it now houses a two-bedroom apartment on the third floor, second floor offices, and on the ground floor a yacht showroom, which is soon to be converted to a restaurant.

The tank's round shape has lent itself to both inside and outside circular balconies. The exterior superstructure provides staircases and access to each floor, as well as great river views; it also provides space for just about the only touch of more common domestic life: planter boxes filled with brilliant pink petunias that complement the building's bright blue exterior skin.

The spaces within circle around a center atrium with balconies. Natural light in the living and office spaces is provided by large windows that have been cut into the tank's walls. As adaptive reuse projects go, this one has obviously taken both imagination and technical expertise, for only its signature shape suggests its former use.

Interior view, with pristine white walls, shows large sliding glass windows, both for natural light and access to the balcony. Only rounded walls suggest the former use of the structure.

LIST OF ARCHITECTS

Anthony Ames
P.O. Box 54144
Atlanta, Georgia 30308
Telephone: 404-266-8904

David Baker
David Baker Associates Architecture
461 Second Street
Suite c127
San Francisco, California 94107
Telephone: 415-896-6700

Philip Banta
Philip Banta and Associates Architecture
6050 Hollis Street
Emeryville, California 94608
Telephone: 510-654-3255

Beyer Blinder Belle
Architects and Planners
41 East 11 Street
New York, New York 10003
Telephone: 212-777-7800

Lawrence Booth
Booth Hansen Associates, Architects/Planners
555 S. Dearborn Street
Chicago, Illinois 60605
Telephone: 312-427-0300

David Burrows
Glueck Architects
116 North Few Street
Madison, Wisconsin 53703
Telephone: 608-251-2551

Clyde Reynolds Carpenter
340 New Street
Gratz Park
Lexington, Kentucky 40507
Telephone: 606-257-3651

Chris Carson
Ford Powell & Carson, Inc.
1138 East Commerce Street
San Antonio, Texas 78205
Telephone: 210-226-1246

Richard Claybour
Claybour Architects
5515 Waterman
Saint Louis, Missouri 63112
Telephone: 314-361-8240

Peter de Bretteville
146 Deepwood Drive
Hamden, Connecticut 06517
Telephone: 203-498-2104

Finegold Alexander
& Associates Inc.
Architects and Preservation Planners
77 North Washington Street
Boston, Massachusetts 02114
Telephone: 617-227-9272

Guy Ladd Frost
5 Sinclair Martin Drive
Roslyn, New York 11576
Telephone: 516-621-2745

Joanne Goldfarb
721 Gibbon Street
Alexandria, Virginia 22314
Telephone: 703-549-7679

Michael Graves
341 Nassau Street
Princeton, New Jersey 08540
Telephone: 609-924-6409

Graham Gund
47 Thorndike Street
Cambridge, Massachusetts 02141
Telephone: 617-577-9600

Michael I. Hauptman
Brawer and Hauptman, Architects
20 North 3rd Street
Philadelphia, Pennsylvania 19106
Telephone: 215: 829-0084

Stephen B. Jacobs
thestephenbjacobsgrouppc
677 Fifth Avenue
New York, New York 10022
Telephone: 212-421-3712

Samuel H. Kaye
P.O. Box 48
Columbus, Mississippi 39703
Telephone: 601-327-6241

Theodore Kessler
481 Park Avenue
Leonia, New Jersey 07605
Telephone: 201-944-3340

Mark A. Kohler
Karl E. Kohler Associates Architects
5206-B Lyngate Court
Burke, Virginia 22015
Telephone: 703-764-7570

Joseph Levine
Bone/Levine Architects
121 Wooster Street, 6F
New York, New York 10012
Telephone: 212-219-1038

Charles W. Moore and Mark Simon
Centerbrook Architects and Planners
67 Main Street
P.O. Box 955
Centerbrook, Connecticut 06409
Telephone: 860-767-0175

Gunnar Orefelt
Orefelt Associates Architecture and Design
Portobello Studios
5 Haydens Place
London W11 1LY, Great Britain
Telephone: 0171-243-3181

OZ Architecture
1580 Lincoln Street
Suite 1200
Denver, Colorado 80203
Telephone: 303-861-5704

Rogers Marvel Architects
145 Hudson Street
Third Floor
New York, New York 10013
Telephone: 212-941-6718

Rothzeid Kaiserman Thomson & Bee Architects and Planners
150 West 22 Street
New York, New York 10011
Telephone: 212-807-9500

Michael Rotondi and Clark Stevens
RoTo Architects
600 Moulton Avenue #305
Los Angeles, California 90031
Telephone: 323-226-1112

Kenneth Schroeder
Schroeder Murchie Laya Associates, Ltd.
936 West Huron Street
Chicago, Illinois 60622
Telephone: 312-829-3355

Josh Schweitzer
Schweitzer BIM Incorporated
5541 West Washington Boulevard
Los Angeles, California 90016
Telephone: 213-936-6163

John Sharratt
John Sharratt Associates
121 Mount Vernon Street
Boston, Massachusetts 02108
Telephone: 617-227-0094

Henry Siegel and Larry Strain
Siegel & Strain Architects
1295 59th Street
Emeryville, California 94608
Telephone: 510-547-8092

Frank G. Stickney
Stickney & Murphy Architects
911 Western Avenue
Suite 200
Seattle, Washington 98104
Telephone: 206-623-1104

Elizabeth Thompson
270 Lafayette Street
Suite 1303
New York, New York 10012
Telephone: 212-274-8336

Stephen Tilly
22 Elm Street
Dobbs Ferry, New York 10522
Telephone: 914-693-8898

Richard C. Tremaglio
5 Story Street
Cambridge, Massachusetts 02138
Telephone: 617-492-4469

Tuck Hinton Architects
410 Elm Street
Nashville, Tennessee 37203
Telephone: 615-254-4100

Troy West
60 Union Street
Newark, New Jersey
Telephone: 973-589-3973

Peter Kurt Woerner
Peter Kurt Woerner and Associates
Studio 44
44 Kendall Street
New Haven, Connecticut 06512
Telephone: 203-466-1923

PICTURE CREDITS

All pictures courtesy of architects and owners, except for the following:

Title Page: Rion Rizzo/Creative Sources
Copyright Page: Mark Luthringer, photographer
Contents Page: Davis A. Gaffga, photographer
Chapter 3: Herb Engelsberg, photographer
Chapter 4: Lloyd Ferriss, photographer
Chapter 6: Norman McGrath, photographer
Chapter 7: Barry Halkin, photographer, also p. 9
Chapter 10: Davis A. Gaffga, photographer, picture 3
Chapter 11: Bruce Van Inwegen, photographer, picture 5
Chapter 12: Steven Rosenthal, photographer, pictures 2, 4, 5
Chapter 15: Craig Schreiner, photographer, picture 2
Chapter 16: James Ritchie and Associates, photographers, pictures 4, 5, 6
Chapter 19: Lisa Mason, photographer, pictures 1, 3, 4, 5
Chapter 20: Alberto Ferrero, photographer, pictures 3, 4, 5
Chapter 21: (all) Rose Island Lighthouse Foundation Collection: Charlotte Johnson, photographer, pictures 2, 4, 5, 6, 7; picture 1: gift of Wanton Chase; Curtis W. Bunting, photographer, pictures 3, 4
Chapter 22: The Sheffield Historical Society, picture 1
Chapter 26: William Kildow, photographer
Chapter 33: Rion Rizzo/ Creative Sources, pictures 3, 5
Chapter 35: Karen Bussolini Photography
Chapter 38: Richard Barnes, photographer, pictures 3, 4, 5
Chapter 39: Toby Abrams, photographer, pictures 1, 3, 4, 5, 6
Chapter 40: Steven Rosenthal, photographer, pictures 1, 4, 5, 6
Chapter 43: Mark Luthringer, photographer
Chapter 46: Thorney Lieberman, photographer
Chapter 50: Phil Mrozinski/M Studio, photographer, pictures 4, 5, 6

Additional Photo Credits and Captions For Front Matter & Filler Photos

Title page: Transformation of a Nashville Shed. Tuck Hinton, Architects

Endsheets: Architect's drawing shows how an old horse barn was reconfigured. Richard C. Tremaglio, Architect

Page 2: "Before" photo of Oklahoma City warehouse: vacant brick building with boarded-up windows, and state of the interior before renovation.

Page 4: Interior of redone Oakland leather factory. Philip Banta and Associates, Architects

Page 7: Retrofitted Greek Revival schoolhouse. Elizabeth Thompson, Architect

Page 8: Library of Michael Graves' converted warehouse.

Page 9: Interior of bank building in Washington, D.C. before renovation by Michael Hauptman, Architect

Page 10: Unusual interior design characterizes updated San Francisco clock tower. David Baker, Architect

Page 11: Artist Laura Warriner made vast live/work space out of an old warehouse interior.

Page 94: West Texas fort, as converted by Chris Carson, Architect.

AMAZING SPACE

Architect Designed Homes Under $300,000 Lucy D. Rosenfeld

Come and explore forty-eight original houses, recently produced by American and Canadian architects for their clients. Pour over the wide variety of innovative house designs represented in over 235 images and plans. While many of these homes are one-of-a-kind works of art, all of the houses presented were produced with an eye for economy. How each budget-minded architect kept the cost low, using a combination of economical construction techniques and clever technological innovations, is discussed and illustrated.

Ranging from starter homes to retirement houses, from vacation retreats to townhouses, from large houses accommodating offices and studios to small residences designed for simple living, these original and economical designs represent a wide variety of architectural outlooks and styles sure to fire the imagination.

8 1/2" x 11" 235+ photos and illus. 160 Pages
hard cover $39.95 0-7643-0642-1

More Schiffer Titles

www.schifferbooks.com

The Virginia House
A Home for Three Hundred Years
Anne M. Faulconer

Lavishly illustrated with over two hundred color photographs, The Virginia House presents a full survey of Tidewater Virginia homes constructed between 1640 and 1830. The buildings range in size from tiny cottages to the great plantation house set in its formal composition of flanking outbuildings and gardens. The emphasis is on the small dwellings, which are affordable yet full of history; and much more suitable for the requirements of twentieth century life. Intricate pictures, floor plans, and details will enable the reader to buy or build his own Virginia dream house, or renovate his current home to project a genteel Virginian image. Personal interviews with the owners of these fascinating homes highlight this engaging, informative volume, which will assuredly expand the popular conception of Virginia homes.

8 1/2" x 11" 200 color plates/50 b/w drawings (HABS) Index
192 Pages soft cover $19.95 0-7643-0598-0

Saltbox and Cape Cod Houses
Revised and Expanded 2nd Edition
Stanley Schuler

From its origins on Cape Cod, Massachusetts in the seventeenth century, this charming and practical style of house has been transplanted and modified to accommodate varying life styles as far away as Hawaii. Stanley Schuler has bought together the architectural history of the Cape Cod House with many floor-plans and photographs to be studied and enjoyed by all who live-in, restore, or want to build their ow Cape Cod House. Examples range from the tiny single style to double, triple, modified and "modern" interpretations-all of which were practical at the time they were built.

8 1/2" x 11" 276 photos Index 160 Pages
hard cover $29.95 0-7643-0998-6

Architectural Details from Old New England Homes
Revised and Expanded 3rd Edition
Stanley Schuler

The doors are wide open and you're welcome to wander through. Don't worry about the carpets as you enter through stately doorways, cozy up to collosal fireplaces, and climb poetic staircases.

For those who love the old homes of the New England area, this is a chance to enter and inspect the window sills and cupboards up close. Nearly 400 photographs and illustrations, along with helpful tips, are provided to guide the remodeller of Jacobean, Colonial, Georgian, and Federal homes toward duplicating these antique architectural features, plus there are architectural drawings from the Library of Congress and by Asher Benjamin, one of the leading New England builders and most influential designers of the late eighteenth and nineteenth centuries.

8 1/2" x 11" 350+ photos Index
144 Pages soft cover $29.95 0-7643-0981-1

Old New England Homes
Revised and Expanded 2nd Edition
Stanley Schuler

Those solid, sensible homes of New England have been studied for their architectural merit and as refections of popular styles in this comprehensive picture book. A wide variety of homes of Colonial, Georgian, Federal and Victorian taste in Maine, New Hampshire, Vermont, Massachusetts, Rhode Island and Connecticut are presented with an informative and readable text.

Floor plans are given for many of the homes, and the 225 color pictures show several angles of the buildings besides frontal views. Historians, architects and home restorers will particularly enjoy this book.

8 1/2" x 11" 438 photos Index 224 Pages
hard cover $35.00 0-76 43-0995-1

Shaker Architecture
Herbert Schiffer

Each of the Shaker communities is represented with very interesting photographs of the buildings that made up their world. The Shakers expressed imagination through their environment, and the details of architecture became their tangible art, while they achieved enduring beauty through the functionalism and simplicity of their design.

8 1/2" x 11" 282 b/w photographs Index 190 Pages
soft cover $24.95 0-88740-153-8

American Barns
Stanley Schuler

American Barns is an exciting, eye-opening study of one of America's most beloved types of buildings. In a handsomely printed text and pictures (113 of them in gorgeous color), it covers barns throughout the United States: huge barns and small barns; Pennsylvania Dutch barns and New England barns; horse barns and carriage houses; and countless others. In all, 240 barns are pictured and explained.

If you love barns, you'll be crazy about American Barns.

If you're thinking about building a barn, American Barns will raise your sights and help you build a better one.

8 1/2" x 11" 281 photos 224 Pages
soft cover $24.95 0-88740-145-7

Schiffer books may be ordered from your local bookstore, or they may be ordered directly from the publisher by writing to:
Schiffer Publishing, Ltd.
4880 Lower Valley Rd
Atglen PA 19310
(610) 593-1777; Fax (610) 593-2002 E-mail: schifferbk@aol.com

Please visit our web site catalog at www.schifferbooks.com or write for a free catalog. Please include $3.95 for shipping and handling for the first two books and 50¢ for each additional book. Free shipping for orders $100 or more.
Printed in China

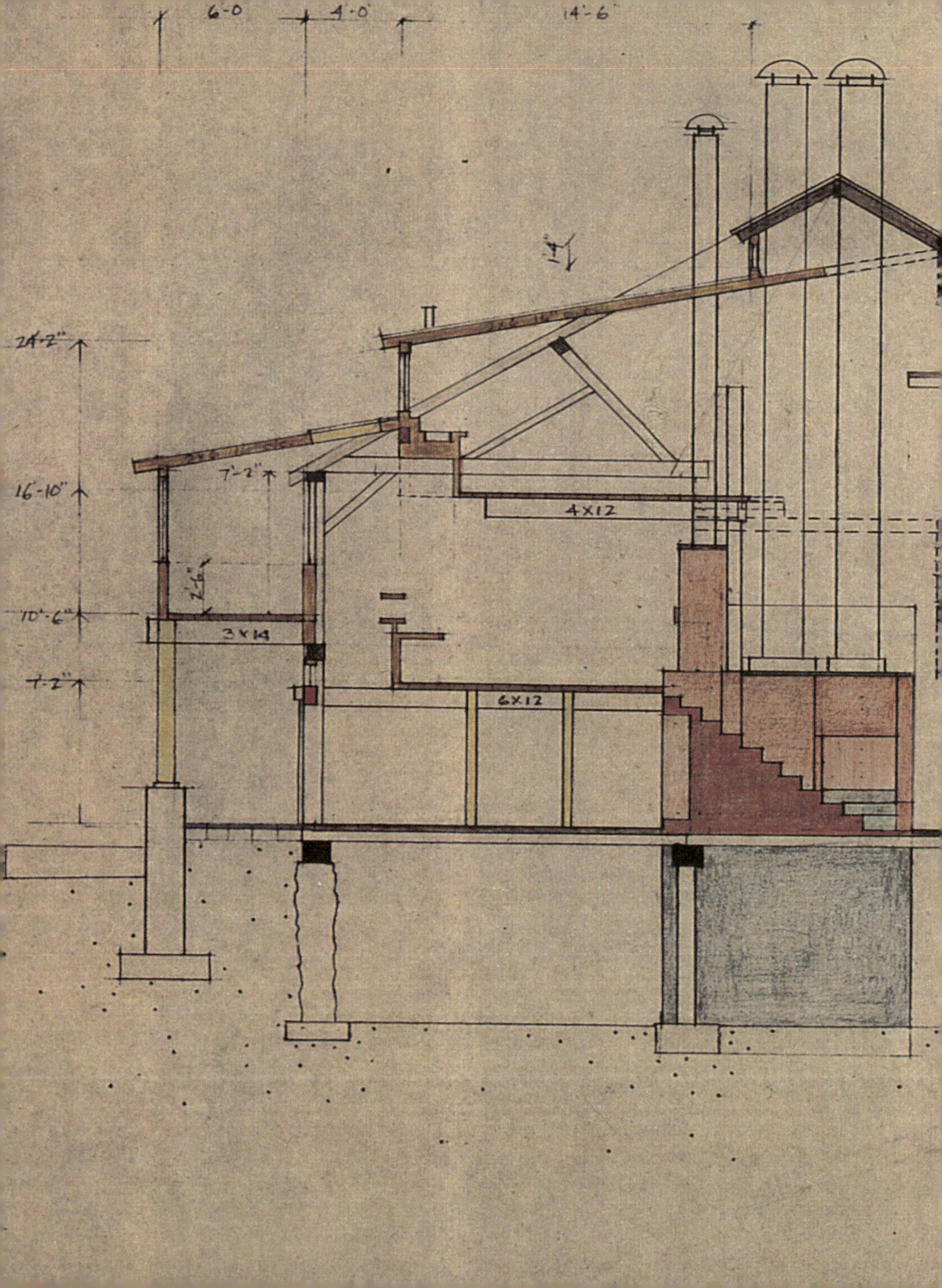
6-0
4-0
14'-6"
24'-2"
16'-10"
10'-6"
7'-2"
7'-2"
2'-6"
3X14
4X12
6X12